THE ROAD TO WAR: WORLD WAR I WESTERN FRONT

A Travel Guide for Exploring the Battlefields of Europe

BY JOHN R. DUNLAVEY

A GLX LLC product

Published in the United States of America

Library of Congress Control Number: 2018905370

ISBN-13: 978-0-692-11831-3

ISBN-10: 0-6921183-1-3

Cover Design by John Dunlavey

Photographs by John Dunlavey, except as noted

Maps by John Dunlavey, except the base maps for France and Benelux, which were completed by Migito Design, LLC.

Front cover photos: Center photo, British Mark V tank with carrier pigeon (NARA); photo ribbon: Newfoundland Memorial, Somme Battlefield, Thiepval Memorial Somme, Meuse Argonne Cemetery, Vimy Ridge Memorial and poppies outside Ypres. Background and base photo: road on westernfront (NARA).

Back cover photos, left to right: Hand grenade found in farmer's field at the Somme 1918, photo ribbon left to right, Allied soldiers, Germans advancing in smoke, young German soldier, French soldiers shot in advance and road on the western front. (NARA)

Dedicated to soldiers from a small town in Pennsylvania who died fighting for others' freedom in the Great War.

PROLOGUE

After America's declaration of war in 1917, many citizens from small towns across America volunteered to fight. The small town of North East, Pennsylvania was no exception. Of North East's approximately 3,500 people in 1915, 237 volunteered to fight, 130 of them deployed to Europe in time for the war, and 26 of those were wounded and 15 killed in action. That is a 33 percent casualty rate, much higher than in other wars. In addition, many of the survivors suffered after the war from war time mustard gas burns, shell shock, and their youth taken from them.

Those killed in action were:

Pvt. Jasper Washburn, killed by sniper fire at Fismette, August 7, 1918; age 18.

Wagoner Victor Carlton Franke, killed by shrapnel while crossing La Selle River, October 17, 1918; age 20.

Pvt. Kenneth Carter, a courier killed by air-burst shrapnel during the battle for St. Quentin, September 28, 1918; age 22.

Pvt. Leslie McGahen, killed in action during 2nd Marne battle, July 29, 1918; age 22.

Pvt.e 1st Class William Simpson, killed in action during the battle for Hill 204, July 7, 1918; age 21.

Pvt. 1 st Class Clifford Sennett, killed in action at Apremont, October 6, 1918 after 95 days of shell fire; age 21.

Salvatore Savaltiere, joined Italian Army in 1915, killed in action fighting the Austrians, July 4, 1916; age 23.

Sgt. Francis Heard, killed in action during the Chateau Thierry battle, July 15, 1918; age 22.

Sgt. James Feeney, killed during training, October 24, 1918; age 25.

Pvt. Leslie Keslar, died of a shell burst while a courier in the Battle for the Argonne, September 29, 1918; age 22.

Sgt Perry Bigelow, Air Service, died in service, of pneumonia, date not noted; age 24.

Pvt. Frank Rundell, died in service, of pneumonia, December 14, 1918; age unknown.

Wayne Benjamin, died in service, of pneumonia, March 24, 1918; age 22.

Cpl. Claire Tickner Hanley, died in service, of pneumonia, December 2, 1917; age 22.

Sgt. Leonard Kightlinger, Aviator, died in service, of pneumonia, March 14, 1919; age 23.

This photo was taken at Camp Hancock, Georgia in 1917. It has some of the boys from my home town including three* who were killed in action. The photo is labeled: Company G and E, 16th Regiment.

Front Row: Lloyd Eaton, Harley Edwards, Earl Darling, Archie Green, Jerome Fileger, Leon Loveless, Harry Wagner, William Mile.

Second Row: Harold Jones, Jasper Washburn*, Charles Lewis, George Allen, Cliff Sennett*, unidentified soldier , Herman Priester, Billy Simpson*.

Back Row: Cecil Chapman, unidentified soldier, James Chapman, Byron Mackintosh, Charter Edwards, C.V. Green, George Baltozer, Roll Chase. Lt. Charles Skellie in the campaign hat, unidentified man.

**Billy Simpson, was stated on this photo as the first soldier from Erie County to be killed in action during the war. The other two were killed later.*

TABLE OF CONTENTS

LIST OF MAPS

LIST OF PHOTOS

TABLE OF CONTENTS

"Though defensive violence will always be a "sad necessity" in the eyes of men of principle, it would be still more unfortunate if wrong-doers should dominate just men."
– Saint Augustine

INTRODUCTION

In 2017, I was inspired to mark the upcoming centennial of the end of the Great War and provide a guide focused on its battles. I decided to update the writings from my last two books and add additional battles to make one dedicated to the Western Front. In doing so, I also wanted to highlight the soldiers and families from my hometown who volunteered and sacrificed much to help defeat the Germans in World War I. In my small rural town, more than 50 percent of the males of fighting age volunteered to fight.

Of those volunteers, some were immigrants, such as Salvatorie Santorini from Italy, who volunteered in 1915 to fight for the Italian Army against the Austrians and died July 4, 1917. In total, 237 men volunteered, of whom 130 crossed the Atlantic and 15 made the ultimate sacrifice. Many more suffered after the war, some dying of complications from chemical attacks, typically of pneumonia or the Spanish flu. Others were scarred for life with undiagnosed or what today is defined as post-traumatic stress disorder. I dedicate this book to those men and their families, many of which would sacrifice again just 25 years later during World War II.

I hope this inspires you to travel to these locations and stop at your town park to read the bronze plaque or stone that memorializes the soldiers and sailors who fought in the Great War.

Please also look for these books in my series:

Volume I – France
Volume II – Belgium, Luxembourg, and The Netherlands

In the near future:

Volume III – Germany
Volume IV – Austria and Hungary
Volume V – Czech Republic, Slovakia and Poland
Volume VI – Ireland
Volume VII - United Kingdom
Volume VIII – Mediterranean (Italy, Greece, Spain, Turkey)
The Road to War – Operation Overlord – Normandy
The Road to War – Operation Dragoon – Southern France to the Vosges
The Road to War – The Battle of the Bulge

I. HOW TO USE THIS BOOK AND TIPS FOR TRAVEL

THE BOOK'S ORGANIZATION:

The book is arranged chronologically, starting with the Siege of Liege in August, 1914 and concluding with the 100 Days Campaign in 1918. For each battle, I provide a description of related events and locations to visit, including GPS coordinates. The "Places to See" also appears on the corresponding maps. Use the coordinates of the "Places to See" to map out your trip and to create your own itinerary based on your location. (You'll discover that many sites are quite close to each other.)

WORLD WAR 1 CONTINENT SUMMARY

MAP 1

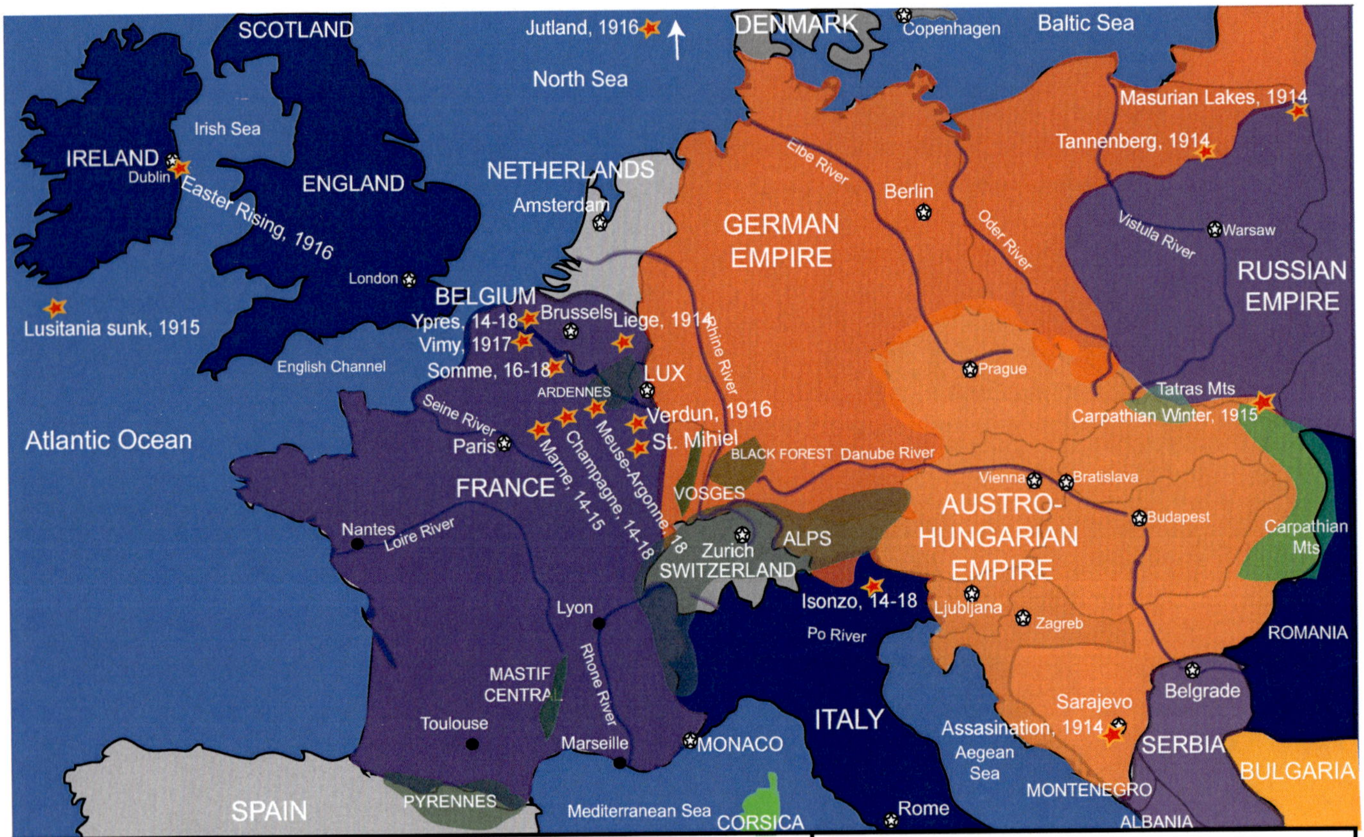

In 1914, the assassination of Franz Ferdinand, in Sarajevo sparked the Great War. This map depicts the borders, major rivers and cities of the Europe, in 1914. Notice, that the German terroritory includes Lorraine and Alsace, from the Franco-Prussian War. Also notice, how much the Ottoman Empire has rescinded off of this map. I only show the major battles and events.

- Allied Powers
- German Empire
- Austrio-Hungarian
- Bulgaria
- Neutral
- Major Battles & Events

WW1 WESTERN FRONT SUMMARY, 1914–1918

MAP 2

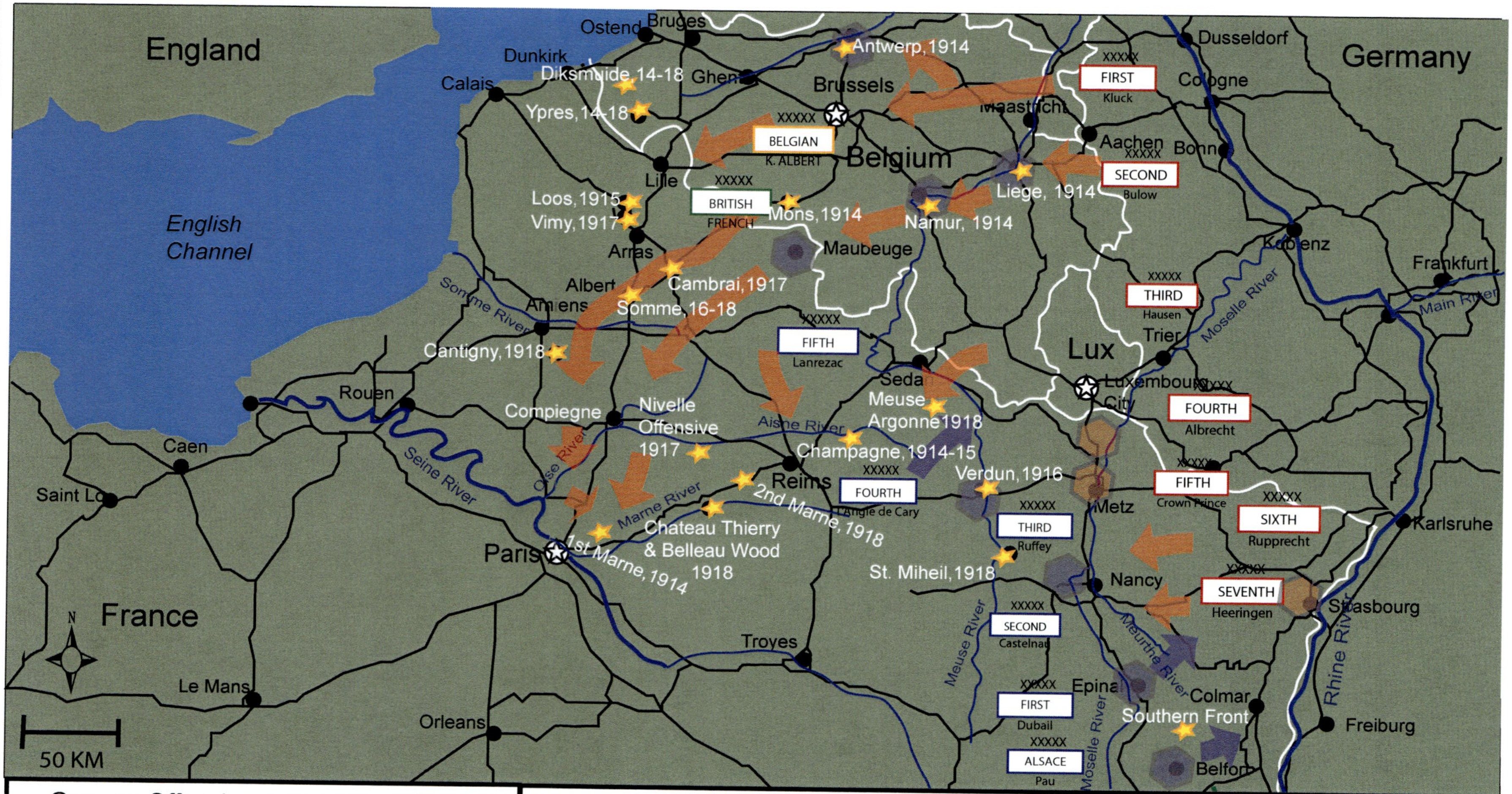

German Offensive, Initial Attack, 1914

- Initial German Attack, 1914
- Initial French Counterattacks, 1914
- Allied Fortified Areas
- German Fortified Areas, keep in mind they still held Alsace since 1870.

This map provides an overview of the initial German attack in August to September 1914. The French responded with a counterattack in the south, which was quickly stopped. The Germans broke through Belgium's main forts of Liege and Namur after almost two weeks of fighting. Then they were blunted at Mons by the British Expeditionary Force (BEF), who pulled back after the French signaled they were withdrawing from Maubeuge Forts. From there the Germans reached the Marne River and fatefully decided to attempt destroying the French Army at the Marne instead of pressing to surround Paris "and failing to "keep the right flank strong" as Schlieffen said to do. At that point from September 5-12 the British and French Armies stopped the Germans at the Marne River and threatend their flanks, forcing the Germans to retreat to Aisne River and dig in. A little less known fact, the Germans under threat in Poland from the Russians, had to pull three Infantry Corps from the west and send them to the east.

WHAT TO EXPECT FOR "PLACES TO SEE":

The following is a quick summary of the types of "Places to See" and what their significance and importance is:

MUSEUMS: The museums include the large national or state-owned museums that always have excellent collections of material and information. I also include many small privately-owned museums that can be a very intimate experience when you engage the owner or caretaker in conversation. Don't hesitate to speak in broken Dutch, French, Flemish, or English; it just adds to your overall experience and you will learn much more. Your interest will also energize them to continue to keep history alive. (It will also entertain them when you don't know French but try to speak it!)

EQUIPMENT: Many "Places to See" are actually pieces of equipment in a town square, on an intersection, or in a field. The equipment ranges from a cannon to a tank. It typically marks an intense battle and the sacrifice that took place there. Usually, they are in memory of a military unit that has spent its own money, together with the town, to refurbish the vehicle or artillery piece to remember the significant action. It is a great opportunity to get up close to the equipment and appreciate the strength and courage it took to ride it into battle. Look at the way it was welded or manufactured, by whom, and examine any battle damage.

STONE MEMORIAL/PLAQUE: Similar to equipment, many monuments have been placed at a bridge, crossing, or at the top of a hill where soldiers fought for victory. They are typically funded through local towns or by local veteran's organizations to remember sacrifices. When visiting these, take your time to get your bearings. Which direction did the attack come from? Where was the approximate breakthrough? Why is the monument at this location?

BATTLEFIELD MARKER OR INFORMATION BOARD: Many of the World War I battlefields have received a "make over" since 1914 because of the centennial event. The markers provide a great opportunity for walking the battlefields.

CEMETERY: Cemeteries are places of mourning, but also a place of honor and remembrance. Individual graves did not start to take place in Europe until WW1. Previously, soldiers were buried in mass graves with no mention of their names. After WW1 and WW2, many smaller cemeteries were consolidated into larger ones and were donated by the host country to the country that made the sacrifice. They are particularly important to visit. Ask the cemetery caretaker to give you a tour and he will tell you some individual stories of sacrifice. Read the names, age, unit, and place they came from. The cemeteries are typically placed in a location of significant battle and sacrifice.

CHURCH: Many places I highlight are churches because they have "seen" a lot of history and for centuries have been the centers of cities, towns, and villages. They are typically at the intersection of centuries-old trade routes from the Roman era. The churches also bear the names of saints who lived in the country and who represent many of the community's values. Their stained-glass windows usually depict events that occurred in the area; some even honor soldiers of WW1. They contain crypts of important leaders, saints, and in many cases, monuments to battles can be found on the inside or outside of the church. During battles, their steeples were used for outlooks for artillery observers. Unfortunately, many churches were scarred by the 80 Years War when they were stripped of their beauty, and then again by the French Revolution when their treasures were stolen and sold as property, and twice more damaged by the World Wars. In any event, today they are still the center of most towns, filled and surrounded with culture and history. Who knows; you could place your hand on the same stone archway that Charlemagne placed his hand on 1,200 years ago, or a wounded soldier of the Great War leaned on, resting with a tin cup of water.

GROUND POINTS: Many battle locations do not have signs and information boards. I find these to be the most interesting and the most rewarding. For their geo coordinates, you need to imagine and visualize the cleared forests, trenches, barbed wire (devil's rope), machine-gun fire, and artillery crashing around. Absorb the terrain, the trees on the ridge, the sloping hills, and the river, and ask questions:

- Did the cavalry hide in there?
- Is it too steep?
- Were the trees really there?
- Was the river wider?

Imagine what the soldier and leaders saw before the attack. Ask questions:

- Where would the best firing positions be?
- Where is the best river crossing point?
- Is this a ditch or is it a trench line and fox hole?

All those things can be discovered in a field or along a river that don't have a monument or battle information board. As Albert Einstein said, "Logic will get you from to A and B, imagination will take you everywhere."

KEY

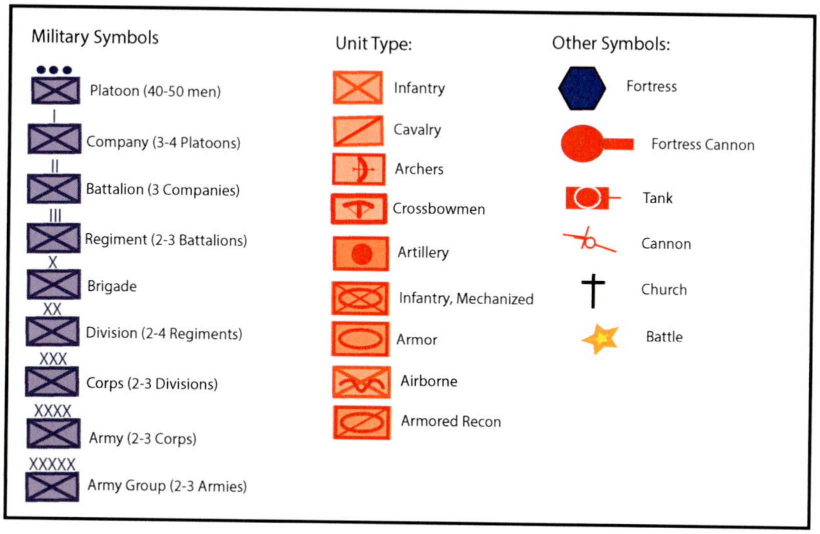

TRAVEL TIPS

Allied troops pull mules through the mud. (NARA)

TRANSPORTATION:

ADDRESS VS. GPS COORDINATES: This book does not use addresses, because based on personal experience, I've found that addresses do not match up well. (Plus, many of the places I included may not have addresses!) For this reason, I only use GPS coordinates. The coordinates and maps are based on WGS84 data. I provide two types of coordinates:

GPS COORDINATE FORMAT

Latitude	Longitude
N 50 06.814	E 8 40.742

DECIMAL AND DEGREES FORMAT

Latitude	Longitude
50.113568	8.679037

Both are for the same location. The majority of rental cars and personal GPS systems use one of those two formats. It is also accepted with the popular phone apps such as Ways and Google Maps.

COMPASS AND MAP: Although you have a GPS, I suggest you buy a European road map and take a compass with you. Many roads are under construction and the GPS may not have all the updates required, or your GPS may malfunction at some point. Either way, navigation can be stressful, so a hard-copy map is very helpful. Having a compass when you are on the battlefield is useful for helping you pinpoint your direction.

ARRIVAL CITY: Because of the volume of airlines arriving daily, good prices are usually available from the USA to Frankfurt, Germany. Using Frankfurt as a hub is also a good idea because you'll easily be able to drive, take the train or connect to other flights. www.Kayak.com is a great tool for finding the best prices for flights. (FYI, Lufthansa has the best service!)

AIR TRAVEL IN EUROPE: This is a great option, as flights are competitively priced in comparison with trains. The planes don't arrive downtown like a train, however! Kayak.com is a great resource for that also. An example for comparison: The trip from Frankfurt to Prague is 8 hours by train, 5 hours driving, or a 45-minute flight.

TRAINS: I personally love taking trains, and Europe has some of the most breathtaking countryside in the world. The downside is that trains can be expensive if you don't book ahead. Always reserve a seat for an additional 5 euros. For train schedules and tickets for all of Europe, check out www.raileurope.com.

RENTAL CARS: Insurance is one of the most expensive aspects of renting a car. Check with your credit card company or insurance company, as they may cover rental insurance, thereby saving you a lot of money. Keep in mind that most rental cars are standard transmission and throughout Europe, many of the cars are diesel. The major rental companies are Hertz, Avis, and Sixt.

SPEED LIMITS AND SAFETY: There are speed cameras everywhere, especially in small towns, so take it slow. In Germany, some areas of the Autobahn do not have speed limits; others do, however, because of congestion, noise abatement, and construction. In most other countries, the speed limit is 130 kilometers per hour. Keep in mind that you should pass only on the left and get to the right when you are done passing. In addition, many US states have agreements with countries for tickets. So, even though you "earned" a ticket in France, they will find you!

WEATHER AND CONSTRUCTION: Typically, Europe does not maintain the same standards regarding snow plows as do the United States or Canada. In other words, if it starts snowing heavily, you may be better off waiting in a town all day for the snow and salt trucks to get some work done. You will also experience road construction. The construction areas can be very tight for passing vehicles.

FUEL: It is expensive. You will pay about $8 to $10 per gallon, or 1.55 euros per liter. Fill up when you can, as you might encounter difficulty finding a fueling station in the more rural parts of the country. Typically, you pump your gas, and then walk inside to pay; in many cases there are no credit-card machines at the pumps. As a courtesy, if the station is busy, you should park your vehicle in a designated area once you have finished fueling, and then go back inside for snacks and to use the restroom.

BORDERS: There is no border control; just be sure to slow down by the empty customs booths when you cross into a different country. Beware: The customs and police do sometimes signal cars to the side for questions.

VIGNETTES: These are toll stickers for your window that you must purchase to drive in Austria and Hungary. They typically cost about 10 euros for three to five days of driving.

TOLLS: As of 2017, the only country with tolls was France. Tolls are used for upkeep, and it is reflected in the country's nice roads and less-congested highways. You can use a credit card at toll booths only if the card has a chip inside it; otherwise, you need cash.

MONEY AND PAPERS, PLEASE:

CASH/CREDIT CARDS: Credit cards are becoming more popular in Europe, but plan on paying cash at many restaurants and the small bed and breakfasts. If they do provide credit card services, however, you sometimes have a choice of charging in US Dollars or Euros. You should always choose to make the transaction in the local currency. If you don't, you will be charged at least 1 percent by your credit card company and the local bank. It can add up, and unfortunately the vendors benefit from it also, so they typically suggest you take the charge in US Dollars. Also American Express is not accepted as widely as Visa is accepted.

VAT: The Value Added Tax is quite high in Europe. It depends which country you visit and what you are buying, but VAT is typically around 18 percent to 20 percent. As a non-EU tourist, you can get the VAT for merchandise back at the end of your journey. If your purchase is more than 40 euros, it is typically worth saving your receipts. Either ask the teller at the store for the form or you can fill in a form with the receipt at the airport prior to your departure.

TIPS: It is best to round up, typically around 7 percent.

ATMS: ATMs are the best way to get cash. Be sure to let your bank know that you will be traveling so your card does not get flagged for fraud (and, consequently, your account does not get frozen). Your bank may charge you a transaction fee when you withdraw cash, but the rate will be better than that at any money exchange.

PASSPORT: You will need a passport on arrival in Europe, and hotels may ask for it to make a copy. When driving across borders in the European Union, you will not need to show your passport.

VISAS: You will not need any visas to travel in the countries discussed in this book.

Commonwealth troops testing communications gear in the trenches. (NARA)

COMMUNICATION:

CELL PHONES: Before your trip, contact your cell phone company to ask for international service, which might include a roaming charge (about 10 dollars per day unlimited use). Or when making phone calls, find a place with free Wi-Fi and use a web service such as Skype, Whats App, or Facetime to call people. Buying a "throwaway" phone with cards is an option, but such phones don't always work in certain countries and the cards run out quickly. Plus, you need to remember a new phone number to share with people who might want to get a hold of you. Therefore, I suggest that you use your own phone.

INTERNET: Many pubs, coffee shops, and chains such as Starbucks and McDonald's have free Wi-Fi.

POWER AND PLUGS: The power in Europe is 220v 50h, as opposed to 110v 60h in the United States. Most laptops and cell phones can use either. Check your charger, which will typically say 110 to 230v and 50 to 60h. While you probably won't need a power converter, you will need a plug adapter. Europe uses a two-prong plug, which you can buy on Amazon for $3.

TRANSPORTATION AND REFERENCES

Left to Right: A main highway in France, typically well maintained and top speeds of 130 KPH. Middle photo is an example of the tight construction on highways in Germany (everywhere). Far right, Europe does not always have the same response or kind of snow plows that we are used to in America, prepare and plan accordingly.

An example of a toll booth in southern France, be ready to use some change or cash. The US credit cards do not always work because our cards sometimes require a signature or pin to process. The middle photo, a good reminder that throughout Europe theft is an issue and something you want to always take precautions for. And apparently there are some dangerous squirrels in Europe too!

The Iron Harvest: Around World War I battlefields the farmers continue to plow up shells and remains from World War I. Be careful and respectful if you venture near fields. This shell was next to a Loos Memorial.

Power and Plugs: The photo to the left is an example of the standard Europe Continent format outlet. The photo right is an example of the simple adapter you can buy for $3 on Amazon or local store in Europe.

II. FOOD, DRINK AND PLACES TO STAY

Commonwealth troops enjoy a short break to eat some grub. (NARA)

FOOD AND DRINK

FOOD AND DRINK: The following provides some of the highlights of food I've enjoyed while traveling in Belgium, France, Luxembourg, and The Netherlands.

FOOD AND DRINK IN FRANCE

As I write this I am starving and very thirsty. Left to Right: Beef Carpaccio, fries and a beer for lunch near St. Raphael. Next photo, I was taking a break with a beautiful view of the Roman Aqueduct, Pont du Gard. Finally, a doner kabab which is always a good and budget smart meal.

In southwest France a typical dish is the Cassoulet, a hearty white bean, duck and sausage stew, gets even better with red wine and fresh bread. Center photo I was sitting admiring the Chartres Cathedral and had an excellent French Onion Soup, fresh bread and a crisp cold coca cola. The photo right, fresh mussels, fries and beer in Normandy, sooo good.

While I was in Arles, I resorted to the traditional French menu. First an appetizer of pate, toast, jam and salad and the house red wine. It was served with fresh bread and bottle of tap water, which in most countries they do not provide you water and bread for free, France does. For my main dish I chose the steak and fries, always a good choice. Finishing a meal with an espresso is always helpful to get you to the next site or enough energy to get you back to your room!

While traveling in Europe and on a budget, I highly recommend getting lunch or snacks at a local grocery store. It is especially helpful on a day you are driving to another location. I typically would buy a fresh baguette, some salami, fresh cheese, bag of chips, some bottles of water and something sweet. In total you can spend about 20 euros to cover 4 people for lunch.

FOOD AVAILABLE IN BELGUIM, LUXEMBOURG AND THE NETHERLANDS

Above left is the traditional Luxembourg dish, "Judd mat Gaardebounen", smoked pork collar that is cooked with beans and potatoes. Upper right, another dish you may see in Luxembourg, an excellent cut of steak, salad and fries.

Available in Luxembourg and Belgium is the above left pate en croute which if fresh is made by the chef of the restaurant. Above right is one of the beautiful streets in Brussels, this one with many excellent options for mussels and fries.

While in the low countries along the coast, you must try the mussels, fries and a beer. The one pictured above left is actually in Bavay, France just across the border from Belgium. Photo above right is a famous french fries stand in Belgium, one of the many that are a great snack for a hungry tourist.

FOOD AVAILABLE IN BELGUIM, LUXEMBOURG AND THE NETHERLANDS

Upper left is the Belgian beef stew with french fries and salad. It is an great option especially on a cold windy day in Bruges! Photo center is another must have in Belgium, a waffle with chocolate syrup and an espresso. Switching to the Netherlands, photo right is the unique "rijsttafel" Indonesian cuisine from the Dutch colonial days. Having indonesian food is a nice break from the traditional European food.

Simple but hearty food can also be found, such as the pea soup above left with some brown bread and a slice of smoked ham. The center photo are the excellent "bitterballen" snack. The snack is usually made with a minced meat with flour and deep fried then served with mustard. Upper right are the other options, the "crouqettes" which contain mashed potatoes and a meat. Photo right shows them with a ham sandwich.

The photo left is of what is always available in the Netherlands... Herring! I can't deny that it tastes great with some white bread and onion and a cold beer. The center photo, as you get closer to Germany, they will have many of the foods we see there such as this boar's stew with pickled cabbage and potatoes. And I can't leave out the excellent "poffertjes" little pancakes with nutella, this one was taken at the Christmas market in Amsterdam's Rembrandt Square.

BEER IN THE BENELUX

Left, a photo of Rochefort Abbey and above, a Rochefort beer. The trappist beers are unique and wonderful beers. The Low Countries, especially Belgium have a wide variety of beers to enjoy in many settings. Below are just some of the many beers you can experience while traveling to these great places.

Above left, a Napoleon beer, brewed at Fort Liefkenshoek, Jupiler light belgian beer, Westmalle another Trappist beer, Hertshogg light Dutch beer. Below left is the enticing Leffe beer sign in Belgium, center a Luxembourg Diekirck light pilsner and signs for the wonderful Orval "golden valley" Trappist beer.

PLACES TO STAY

French troops in a dugout, in relative safety rest, and the thousand-yard stare. (NARA)

HOTELS AND B&BS: There are many great hotels and B&Bs, but also some bad ones. I highly recommend using www.Tripadvisor.com or www.Booking.com to find the best price range for your accommodation needs. Note that breakfast at hotels may be expensive (20 euros or so). You can opt out of that by asking the hotel not to include breakfast during your stay. Also consider calling the hotels to book versus using the websites; they may have better pricing and offer different rooms.

Following is a list of places to stay in cities and strategic locations near the World War I battlefields in Belgium and France.

MAP 3

PLACES TO STAY

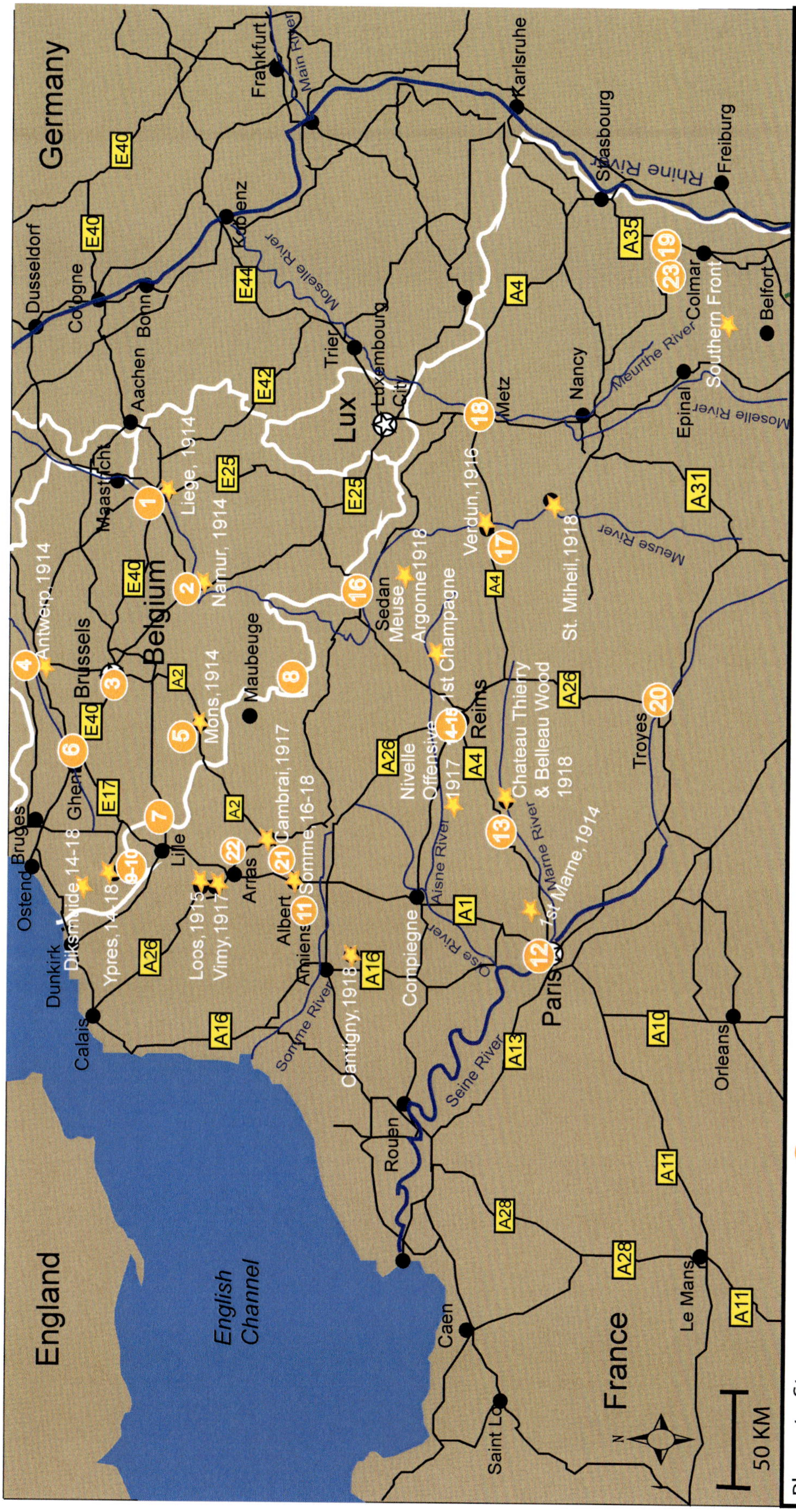

Places to Stay:

1. Liege, Penta Hotel
2. Namur, Chateau de Namur
3. Brussels, Stephanie Hotel
4. Antwerp, Leonardo Hotel
5. Mons, B&B
6. Ghent River Hotel
7. Tournai, B&B
8. Couvin Chateau
9. Ypres, B&B
10. Ypres, Main Street Hotel
11. Somme, Hotel
12. Paris, Hotel Sainte Anne
13. Chateau B&B
14. Reims B&B
15. Reims B&B
16. Sedan Chateau
17. Verdun Hotel
18. Metz B&B
19. Zellenberg Hotel
20. Troyes, Le Jardin
21. Cambrai Hotel
22. Arras Hotel, Hotel la Ferme Sensee
23. Riquewihr Hotel, Couronne Hotel

1. **LIEGE, PENTA HOTEL:** A very modern "edgy" hotel, but excellent and very affordable.
www.pentahotels.com

100 Blvd de la Sauveniere, 4000 Liege, Belgium
Tel: +32 011 32 42 68 22 77
Coordinates: N 50 38.567 E 5 33.922 (50.642786, 5.565368)

2. **NAMUR, CHATEAU DE NAMUR:** A beautiful old hotel, located next to the Namur Citadel; an excellent place to stay.
www.Chateaudenamur.com

1 Avenue de l'Ermitage, 5000 Namur, Belgium
Tel: +32 81 72 99 00
Coordinates: N 50 27.288 E 4 51.334 (50.454794, 4.855559)

3. **BRUSSELS THON BRISTOL STEPHANIE:** An affordable four-star hotel in an excellent area of Brussels.
www.thonhotels.be

91-93 Avenue Louise, 1050 Brussels, Belgium
Tel: +32 25 43 33 11
Coordinates: N 50 49.887 E 4 21.538 (50.831456, 4.358971)

4. **ANTWERP, LEONARDO HOTEL:** A modern and excellent hotel in central Antwerp, close to all the major sites.
www.leonardo-hotels.com

59 De Keyserlei 59, 2018 Antwerp, Belgium
Tel: +32 32 32 14 43
Coordinates: N 51 13.038 E 4 25.211 (51.217300, 4.420184)

5. **MONS, COMPAGNONS11 B&B:** Located only a five-minute walk from the city center, a modern and well-priced B&B, especially given the amenities.
www.compagnons11.be

11 Rue des Compagnons, Mons 7000, Belgium
Tel: +32 499 99 76 60
Coordinates: N 50 27.378 E 3 56.908 (50.456306, 3.948472)

6. **GHENT, RIVER HOTEL:** An excellent location, right downtown near the castle and canals. A great hotel and city. For parking, there is street parking and a public parking lot. The breakfast area view is spectacular.
www.ghent-river-hotel.be

5 Waaistraat, 9000 Ghent, Belgium
Tel: +32 92 66 10 10
Coordinates: N 51 03.491 E 3 43.561(51.058190, 3.726023)

7. **TOURNAI, LA ROSERAIE B&B:** Approximately 8 km outside of Tournai toward the French border, and in the country. It is a beautiful and very accommodating B&B, a great location for a point between the main actions, especially WW1 battlefields.
www.laroseraie-tournai.com

113 Rue de Roubaix Templeuve, Tournai 7520, Belgium
Tel: +32 (0) 69 36 02 76
Coordinates: N 50 38.686 E 3 16.273 (50.64476 3.271219)

8. **COUVIN, CHATEAU TROMCOURT:** In the country and a beautiful setting with a welcoming staff and affordable accommodations.
www.tromcourt.com

15 Chemin de Tromcourt, 5660 Couvin, Belgium
Tel: +32 60 31 18 70
Coordinates: N 50 05.980 E 4 29.537 (50.099663, 4.492289)

9. **YPRES, VARLET FARMS B&B (COUNTRYSIDE):** This B&B sets the benchmark for B&Bs near battlefields. When I visited in 2011, I was greeted with an apple pie, coffee, and a quick intro to the area. It continues to be highly ranked and is close to many of the battle sites.
www.varletfarm.com

43 Wallemolenstraat, 8920 Poelkapelle, Belgium
Tel: +32 51 77 78 59
Coordinates: N 50 54.346 E 2 59.061 (50.905763, 2.984349)

10. **YPRES, MAIN STREET HOTEL (DOWNTOWN):** An excellent hotel located in downtown Ypres. It is nice to have the option of going out to dinner in beautiful Ypres and then conveniently walking back to the hotel.
www.mainstreet-hotel.be

136 Rijselstraat, 8920 Ieper, Belgium
Tel: +32 57 46 96 33
Coordinates: N 50 50.797 E 2 53.352 (50.846623, 2.889200)

11. **SOMME, ALBERT: BEST WESTERN ROYAL PICARDIE.** Although the hotel is located outside of Albert, it is very convenient and allows for easy exploration of all the Somme locations.
www.royalpicardie.com/en/

138 Avenue du Général Leclerc, 80300 Albert
Tel: +33 3 22 75 37 00
Coordinates: N 49 59.904 E 2 38.095 (49.9984, 2.634917)

12. **PARIS: HOTEL LOUVRE SAINTE ANNE.** Located within walking distance of the Louvre and Opera House, this hotel is also conveniently located near a secure public-parking garage.
www.paris-hotel-louvre.com/

32, Rue Sainte-Anne, 75001 Paris
Tel: +33 1 40 20 02 35
Coordinates: N 48 51.999 E 2 20.147 (48.86665, 2.335783)

13. **CHÂTEAU THIERRY: CHAMPAGNE B&B.** This B&B is a fantastic place to stay. It is owned by American expatriates who are very knowledgeable about the area and places to visit. Halfway between Paris and Reims, it is very convenient for seeing the area's World War I battle sites.
www.marneweb.com

1 Rue des Vaches, 02850 Reuilly-Sauvigny
Tel: +33 6 13 24 46 16
Coordinates: N 49 03.486 E 3 33.393 (49.0581, 3.55655)

14. **REIMS: MAISON D'HOTES LES TELLIERS.** Located in the city center near the cathedral, this B&B is affordable and convenient. You must coordinate parking with the hotel, however.
www.telliers.fr/

18 Rue des Telliers, 51100 Reims
Tel: +33 9 53 79 80 74
Coordinates: N 49 15.397 E 4 01.752 (49.256617, 4.0292)

15. **REIMS: LE CLOS DU MONT D'HOR.** Located outside Reims, this is a beautiful place to stay, among the vineyards and the champagne!
www.mhchampagne.com/en/

8 Rue du Mont d'Hor, 51220 Saint-Thierry
Tel: +33 3 26 03 12 42
Coordinates: N 49 18.334 E 3 58.291 (49.305567, 3.971517)

16. **SEDAN: HOTELLERIE LE CHÂTEAU FORT.** This hotel is located inside the very impressive castle of Sedan. When you arrive at the castle, to enter the hotel you will have to drive through the main gate and park on the right.
www.Chateaufort-sedan.fr/en/

Porte des Princes, 08200 Sedan
Tel: +33 3 24 26 11 00
Coordinates : N 49 42.148 E 4 57.006 (49.702474, 4.950092)

17. **VERDUN: HOSTELLERIE DU COQ HARDI.** This hotel is located in the small downtown of Verdun and is convenient for exploring the area.
www.coq-hardi.com/

8 Avenue de la Victoire, 55100 Verdun
Tel: +33 3 29 86 36 36
Coordinates: N 49 09.673 E 5 23.142 (49.161216, 5.385693)

18. **METZ: LES CHAMBRES DE L'ILE B&B.** Located downtown within walking distance of the cathedral, this hotel is also close to many locations of interest.
www.bnbmetz.com/

15 Rue de l'Horticulture, 57050 Longeville-lès-Metz
Tel: +33 6 13 23 28 33
Coordinates: N 49 06.863 E 6 09.432 (49.114389, 6.157192)

19. **ZELLENBERG: LE SCHLOSSBERG.** The town of Zellenberg is a convenient location for visiting the Alsace Plain between Colmar and Selestat. This area has many great B&Bs and hotels. Le Schlossberg is a very welcoming, quiet, and clean B&B with updated amenities.
www.leschlossberg.com/en/

59a Rue Fontaine, 68340 Zellenberg
Tel: +33 3 89 47 93 85
Coordinates: N 48 10.080 E 7 19.133 (48.168, 7.318883)

20. **TROYES: LA JARDIN DE LA CATHEDRALE.** Located in downtown Troyes within walking distance of shopping venues and attractions.
www.jardindelacathedrale.com

12 Place Saint-Pierre, 10000 Troyes
Tel: 33 6 63 10 32 32
Coordinates: N 47 19.088 E 5 02.206 (47.318126, 5.036771)

21. **CAMBRAI: HOTEL BEATUS**. Located just outside of Cambrai, a beautiful small hotel, great to explore the region, owned by the Philippe Gorczynski, the tank researcher.
www.beatus-cambrai.com

718 Avenue de Paris, 59400 Cambrai
Tel: +33 3 27 81 45 70
Coordinates: N 50 09.837 E 3 13.684 (50.163954, 3.228069)

22. **ARRAS: HOTEL LA FERME SENSEE.** This beautiful and quiet horse farm is located in the town of Gouy-Sous Bellenne, about 20 minutes outside Arras. It is a great location and wonderful for getting away from the larger cities. **www.fermedelasensee.free.fr**

    5 Rue de l'Église, 62112 Gouy-Sous-Bellonne
    Tel: +33 3 21 50 47 31
    Coordinates: N 50 18.732 E 3 03.297 (50.312192, 3.054947)

23. **ALSACE, RIQUWIHR, HOTEL COURONNE:** A great 1500s-built structure turned into a comfortable hotel with character and all the modern comforts. **www.hoteldelacouronne.com**

    5 Rue de la Couronne, 68340 Riquewihr, France
    Tel: +33 3 89 49 03 03
    Coordinates: N 48 09.985 E 7 17.870 (48.166415, 7.297826)

World War I

1914–1918

The war to end all wars was a tragic and terrible conflict that involved new technologies for death and destruction that resulted in never-before seen numbers of casualties. Furthermore, it left behind a broken world that fostered the rise of such radicals as Francisco Franco, Benito Mussolini, Joseph Stalin, and Adolf Hitler. World War I ended with 22,477,000 military casualties for the Allies and 16,403,000 casualties for the Central Powers. This book highlights the major Western Front battles, as well as dozens of museums, cemeteries, and memorials. The region has a wealth of locations to visit, learn about, and honor. The following is a quick summary of what happened to France, Belgium, the Netherlands, and Luxembourg during that war, and is followed by a timeline.

World War I was ignited when on a tour of Sarajevo, Archduke Franz Ferdinand of Austria and his wife Sophie, Duchess of Hohenberg, were assassinated by a Serbian nationalist, Gavrilo Princip of The Black Hand, on June 28, 1914. In response, a month later on July 26, the Austrian Hungarian Empire declared war on Serbia. Then bound by treaty as an ally of Serbia, Russia declared war on Austria-Hungary. The chain reaction accelerated; France, which was bound by treaty to Russia, mobilized her military. In response, Germany, an ally of Austria-Hungary, began to mobilize and declared war on France, Russia, and Japan on August 1.

On August 1, the Germans initiated their Schlieffen Plan to invade France by surprise. They aimed to attack quickly through neutral Belgium and Luxembourg, with the goal of surrounding Paris and ending the war quickly. On August 2, the Germans requested approval to pass through Belgium. The Belgium government refused on August 3. In response, the Germans invaded Belgium on the 4th. The Belgium forces fought bravely against an overwhelming German army, and because of their courage the German advance was stalled. That gave the British and French more time to assemble a defense. As described in this chapter, some of the bloodiest fighting took place in Belgium, around the Liege forts, Mons, and Ypres. The Belgian civilians also suffered terribly under German occupation. The Germans killed more than 5,000 civilians for perceived "subversion" and espionage, from August to November, 1914.

While fighting through Belgium, the Germans were also fighting in the east against Russia in what is Poland today. The battle in the east required an additional three German corps (circa 100,000 soldiers) to stave off a loss to the Russians. After the Germans received the additional troops in the east, they destroyed the Russian army in August at Tannenburg. The Grand Duke Nicholai, Chief of Staff of the Russian Army, said after the devastating loss, "It is an honor to make such sacrifice for our Allies." He said that because the corps were pulled from the western offensive, which weakened the overall Schlieffen Plan "sickle." He was suggesting that was why the Germans did not reach Paris in September, 1914, and he may have been right.

NOTE ON LUXEMBOURG AND THE NETHERLANDS: In 1914, the Grand Duchy of Luxembourg was a nation of 266,000 people with an army of 400 soldiers. The country was neutral when the German's Schlieffen Plan offensive began on August 1. Initially, the Germans only crossed over the far north of Luxembourg, but then they passed through the south. On August 2, they declared that they needed to occupy the country, but only until their military goals in France were completed. As history tells us, the occupation continued until 1918. Of the Luxembourg citizens who lived overseas; more than 3,700 joined the French army and of those, 2,000 died fighting for freedom.

Like Belgium and Luxembourg, the Dutch were also neutral. Due to their location, however, they were never occupied by either the Allies or the Central Powers. The Netherlands was inundated with Belgian refugees and became a center for food, raw materials, and transportation for both sides of the war. Like other neutral countries, it also became a center for espionage, as people with Dutch passports were allowed to travel throughout all of Europe, including the famous Dutch dancer and double spy Mata Hari. She was eventually caught by the French in 1917 and executed.

Archduke and wife leaving the Sarajevo Town Hall only minutes before their assassination. (NARA)

1914

June 28:	Archduke Franz Ferdinand of Austria and his wife Sophie, Duchess of Hohenberg are assassinated by a Serbian nationalist, Gavrilo Princip.
July 26-28:	Austria declares war on Serbia. As a Serbian ally, Russia mobilizes to declare war on Austria; and it calls on the French to mobilize. On August 1, 1914, Germany declares war on Russia because Germany is an ally of Austria.
August 1:	Germany begins to mobilize its army while declaring war on France.
August 4:	The Germans advance into Belgium and Luxembourg.
August 4:	England and Canada declare war on Germany after Germany invades and does not withdraw from Belgium.
August 5-16:	**Battle of Liege**
August 23-30:	During the Battle of Tannenberg, Russia invades Germany and Germany crushes the Russian First Army.
September 9-14:	During the Battle of Masurian Lakes, the Germans partially destroy the Russian Second Army.
September 5-10:	The **First Battle of the Marne** ends in a French victory and halts the German advance.
September 14:	During the **First Battle of Aisne,** troops on both sides begin to construct trenches.
October 14-November 22:	The **First Battle of Ypres** ends with no decisive victory.
October 29:	Turkey enters the war on Germany's side.

1915

January 1-March 30:	The first Allied offensive in Artois and Champagne ends with no decisive results.
January 19:	The first German zeppelin air raid on England takes place.
April 22-May 5:	During the **Second Battle of Ypres,** the Germans use poison gas for the first time.

April 23-January 9, 1916:	The British and Allies attack Turkey at the Dardanelles and Gallipoli.
May 7:	A German U-boat sinks the British liner Lusitania, which has 128 Americans on board.
May 23:	Italy enters the war on the side of the Allies.
June 29:	Italy launches an unsuccessful attack against Austria-Hungary at Isonzo; there will be a total of 12 attacks throughout the war.
September 22-November 6:	During the **Second Battle of Champagne,** the Allied offensive ends with 145,000 casualties and no ground gained.

1916

February 22-December 18:	The Germans launch a surprise attack at **Verdun,** which becomes the longest battle of the war. It lasts 7 months, leads to 700,000 casualties, and results in no territory gained.
March 9:	Pancho Villa raids Columbus, NM.
April 24:	The Irish Republican Easter Rebellion begins; it is a failed battle against British occupation.
May 19:	Britain and France conclude the Sykes-Picot agreement to divide the Ottoman Empire after the war.
May 31-June 1:	The Battle of Jutland ends without a clear victor. It is the biggest naval battle in history.
July 1-November 13:	In the **Battle of the Somme,** England suffers 60,000 casualties on the first day of fighting.
July 19-20:	**Battle of Fromelles**
September 15:	Tanks are used for the first time, by the British in the Somme Offensive.
September 20:	The Germans began building **The Hindenburg Line.**

1917

February 24:	The Zimmerman Note is discovered and passed on to the United States; the telegram details a German request that Mexico attack the United States.

March 11:	British troops take Baghdad from the Ottoman Turks.
March 16:	**Operation Alberich,** the Germans withdraw to the Siegfried Stellung portion of The Hindenburg Line.
April 10-12:	The Canadians gallantly take Vimy Ridge as part of the British diversionary attack on Arras.
April 11 and May 3-17:	**Battles of Bullecourt**
April 16-May 9:	The French Nivelle Offensive, which includes the **Second Battle of Aisne** and the **Third Battle of Champagne,** ends in a French defeat. Only 457 meters are gained in exchange for 250,000 casualties.
June 4:	The United States declares war on Germany.
June 7-14:	**The Battle of Messines** begins. It is a British offensive and the prelude to the Third Battle of Ypres (known as Passchendaele). In this battle, the British detonate 19 mines under the German lines.
July 6:	T.E. Lawrence (Lawrence of Arabia) leads the Arabs and captures Aqaba from the Ottoman Turks.
July 16–November 6:	The **Third Battle of Ypres** takes place. It is a British offensive that reaches Passchendaele in Belgium, but proves to be strategically inconclusive, with approximately 250,000 casualties on both sides.
November 2:	The British government releases the Balfour Declaration; it commits them to establishment of a Jewish state and details a postwar plan to divide most of the Middle East and Africa.
November 7:	Bolsheviks overthrow the Russian government and install Communist rule under Vladimir Lenin (after the Germans had helped bring Lenin back from exile in Switzerland to Russia).
November 20-December 7:	The **Battle of Cambrai**
December 3:	The German/Russian armistice is signed.
December 9:	The British take Jerusalem from the Turks.

1918

March 21-April 5:	**Battle of Picardy,** Operation Michel and the German Spring Ludendorff Offensives begin.
April 9-29:	**Battle of Lys,** Operation Georgette, the second major German Spring Offensive
April 24-27:	**Battle of Villers-Bretonneux**
May 27-June 6:	The **Third Battle of Aisne,** Operation Blucher and the 3rd German Offensive attack.
May 27:	The United States wins its first major action, at the **Battle of Cantigny.**
June 8-12:	**Noyon Battle,** Operation Gneissenau, the fourth German Spring Offensive attack
June 1-26:	**Battle of Belleau Wood**
July 4:	**Battle of Hamel** outside of Amiens
July 15:	The last German attack, **"Second Battle of the Marne,"** Operation Friedensturm
July 18-August 6:	Allied counterattack at the Marne Salient
August 8:	The **100 Days Offensive begins,** and during the Amiens Offensive, the Allied forces counterattack and force the Germans back to the Hindenburg Line.
September 12-18:	**St. Mihiel Offensive,** largely American forces recapture the German salient south of Verdun.
September 26-October 28:	**Champagne Offensive,** American and French forces push the Germans back.
September 29-October 2:	**Battle for St. Quentin Canal;** Americans, British, and Australians breach the Hindenburg Line.
September 26-November 11:	The **Meuse Argonne Offensive** threatens Sedan, helps push the Germans to the Armistice.
November 11, 1918:	Fighting ceases at the 11th hour on the 11th day of the 11th month of 1918.

THE SIEGE OF LIEGE AND BELGIAN FORTS

August 5-16, 1914

"Frayez-vous le passage" (you must fight your way through)
—Liege fort commander to the Germans' envoy, August 1914.

SUMMARY: On August 1, the Germans initiated the Schlieffen Plan to invade France and on August 2, they requested permission to travel through neutral Belgium. On August 3, the Belgians refused. On August 4, the Germans marched into Belgium and after some skirmishing, they reached Liege. Thus began the first major fighting of the war. The Belgians fought outgunned and outnumbered but delayed the Germans enough to allow the French and British to set up their defenses.

For centuries the Belgians and their rulers had built forts to guard the major thoroughfares and ports. In the mid-1800s when Napoleon III came to power in France, the Belgians who had recently won their independence in 1830, decided on a strategy of a "national redoubt." They faced the real possibility of the French and Germans fighting once again and attacking through Belgium. The strategy called for building strongholds at Liege, Namur, and Antwerp for the Belgian forces to retreat to. They aimed to keep their positions in the fortresses until their Allies, such as the British, arrived to help.

To be ready for the defensive, the Belgians decided to upgrade their forts, such as those around Antwerp where they built a second ring of forts. Under several governmental decisions, the Acts of 1851, 1859, 1870, and 1906, they continued to upgrade and add forts because of the ever-changing nature of the threats against them and the new technologies. One, for example, was the more powerful cannons with longer ranges. Prior to WW1, the Belgians realized their forts were too close to Antwerp, so they pushed the next ring farther out. They provisioned the forts with longer-range cannons so they could support each other, and they reinforced them to withstand rounds from 210mm cannons. Before WW1, however, the German company Krupp had built "Big Bertha," a 420mm siege cannon, and the Austrian company Skoda produced a 305mm cannon. (At that time, Skoda was located in Bohemia in the Austrian Empire, which later became the modern Czech Republic.) In addition, the forts were woefully inadequate in terms of ventilation, septic tanks, and water. Also, the cannons fired black-powder charges, not smokeless rounds, creating a terrible environment for the defenders.

As the crisis escalated and war seemed imminent, the Belgians decided that if attacked they would send a division to Liege, one to Namur, and the rest of the army to Antwerp, the "National Redoubt." Thus, when the Germans advanced the Belgians stationed the 3rd Infantry Division (ID) in Liege, together with four brigades totaling

30,000 men and accompanied by the garrison of 6,000 men under the command of Lt. General Gerard Leman. The Belgians sent the 4th ID to the forts of Namur. For the attack, the Germans formed the Army of the Meuse, consisting of 6 infantry brigades (11th, 14th, 27th, 34th, 38th, and 43rd Brigades) and three cavalry divisions (9th, 2nd, and 4th), for a total of 59,600 men and 100 cannons, under the command of General Otto von Emmich.

On August 4, the Germans initially sent aircraft over Liege to drop pamphlets. Next, as the sun rose, the brigades arrived at their jumping-off points. The German infantry received some resistance from the various forts. The 34th Infantry Brigade (IB) approached on the right flank (north) with the intention of crossing the river at Vise. The 27th, 14th, and 11th Brigades advanced in the center across from Liege. The 38th and 43rd Brigades advanced southeast of Liege and the 9th and 2nd Cavalry Divisions (CDs) captured crossings at Poulseur over the Ourthe River, south of Liege. Thus, the Germans were set, but the element of surprise was gone. They sent an envoy to offer terms, but Belgian Commander Leman rejected them. He then sent infantry out to defend and skirmish to the south and east of Liege.

The German plan of attack called for the 34th IB and 4th Cavalry to attack between Fort Loncin and Fort Pontisse in the north, the 27th IB past Fort Evegnee, the 14th between Fort Evegnee and Fort Fleron, the 11th past Fort Fleron and Fort Chaudfontaine, the 38th and 43rd between the Ourthe and Meuse Rivers, and the II Cavalry Corps (CC) to assemble to the northwest of Liege to block Allied reinforcements.

On August 6 at 0230, the Germans began their assaults. By dawn in the north, the 34th IB had reached the high ground near Liege, but by 1030 they had to retreat because of bombardments from the forts. The 27th IB in the center didn't have much more luck and was held up at Blegny by the fort's defensive fire. The results were the same for the 11th Brigade in the center; it attacked Magnee and Hadelin, but eventually had to retreat because of shell fire. The 38th and 43rd also made some advances, but then retreated under fire from the forts. The 14th Brigade in the center, however, had captured the high ground near Liege at the village of Queue du Bois, together with the citadel. That was followed by a gamble, to rush the bridges into eastern Liege, which proved a smart and courageous move for the Germans. By noon, the Germans were in Liege. In response, Leman ordered the 3rd ID to depart Liege via the northwest to Gete, halfway to Antwerp, and he moved his HQ to Fort Loncin, the westernmost fort around Liege.

Of significance, at 0300 on August 6, a zeppelin dropped bombs on Liege for the first air attack in history on a European city, but the attack had little effect and the airship crash-landed near Bonn after being damaged by ground fire.

On August 8, the next phase of the Siege of Liege began with the arrival of the VII, IX, X Infantry Corps under the command of General Karl von Einem. He brought heavy artillery with him, including the 305mm and the 420mm "Big Bertha" siege guns. The Germans set up the cannons inside the ring of forts, allowing them to shell the forts from behind and systematically pounded them. When engaged with a fort, they fired 200 to 300 rounds per hour (various calibers), and after a few hours the forts surrendered one by one.

Fort Barchon fell first, on August 8, then Fort Evegnee was captured. By August 12, the 305mm and 420mm cannons were in full action. That day, Fort Chaudfontaine and Fort Pontisse fell, on August 14, Fort Liers and Fleron surrendered. On August 15, the Germans began firing on the western ring forts. That same day, Forts Boncelles and Lantin fell and Fort Loncin was destroyed instantly when a round hit an ammunition bunker. Commander Leman, who was in Fort Loncin was wounded, knocked unconscious and then captured. Liege was in German hands.

The Germans continued to advance from August 20-24, laying siege to the fortified position of Namur. This time, the Germans decided to hold off on an infantry attack and just wait for the siege cannons. During the siege, the Germans defeated a French relief force at Charleroi on August 21. Then the siege cannons made short order of the "new" Namur forts; by August 24, all had surrendered. The Belgians suffered 15,000 casualties and the Germans almost 1,000.

Following Namur, the Germans faced the British at Mons in a battle that will be described in the next section. After defeating the British (some would say the British withdrew), the Germans fought the French at Fort Mauberge for a limited time. After advancing deeper into France, the Germans returned to besiege Antwerp between September 28 and October 10. On September 30, they destroyed Forts Walem, Sint Katelijine, and Waver. Then, on October 2, the Germans destroyed Fort Lier, October 4 Fort Kessel, and finally Fort Broeohen on October 6. On October 9, the Belgian and British relief forces abandoned the right bank of the Scheldt River. On their departure, the Belgians blew up Forts Schoten and Brasscharat. On October 10, the Belgians pulled back to the Yser Line with approximately 80,000 men, and held that line for the rest of the war.

OUTCOME: The result was severe bleeding of the German main force and at least a 5-day delay in the Schlieffen Plan. The Germans had believed the siege of Liege would only take 2 days, but the courage of the Belgians was successful in providing additional time for the British and French to mobilize.

WHO? NUMBERS AND LEADERS? Liege was defended by the 3rd Division and fortress troops under the command of Lt. Gen. Gerard Leman, a total of 36,000 men and 252 cannons in Liege, and they suffered 20,000 casualties. The Germans attacked Liege with the Army of the Meuse and 59,800 men under the command of Gen. Von Emmich. They suffered 5,800 casualties in the first 3 days alone; there are no records for after that time.

WHEN? August 4, 1914.

WHY AT LIEGE? Liege was on the main line of attack for the Germans and was the Belgian's first defensive fort. It sits at the confluence of the Meuse River and on the railroad line from Liege to Brussels to Paris.

TECHNOLOGY/TACTICS: The Belgians built "modern" forts in 1888-1891. They decided to build them in Liege and Namur because they were in the direct path between France and Germany. After the Franco-Prussian war, the Belgians hoped to dissuade the Germans and French from attacking each other again. The forts were built using the "new" material of concrete, but reinforced rebar had not yet been invented. Nevertheless, they made the roofs 4 meters thick so they could stop shells from 210mm cannons. Interestingly, the Belgian cannons were made by the German company Krupp and some of the turrets were made by the French and Germans. Unfortunately, although "modern," the forts still lacked forced air, did not have proper latrines, telephone lines were above ground, the cannons fired black-powder charges, and they had insufficient water supplies.

The 12 Liege forts were built in a 6 to 10 km ring around the city. The forts had ditches, 75mm, 150mm, and 120mm cannons, and 57mm quick-loading cannons. They were built to defend Liege, but if Liege were entered, they could be attacked from the rear. As mentioned, the Belgian tactic was to delay the Germans as long as possible at Liege and Namur while the main force retreated to Antwerp to wait for the Allies to assist. The Germans, on the other hand, attacked Liege with infantry, although at first unsuccessfully. Not until they brought up the heavy artillery could they breach the forts. They did so with the new Krupp-built 420 mm "Big Bertha" cannons and the Austrian Skoda-built 305mm cannons. During the sieges, the Germans fired 590 shells from the 420mm cannon and 2,130 from the 305mm cannon. They then adjusted their tactics when they attacked Namur, first bringing in the howitzers, then attacking with infantry.

PLACES TO SEE:

FORTIFIED POSITION OF LIEGE: Following are the forts around Liege and additional places to visit. The forts' construction started in 1888. Some were upgraded and integrated into the defensive plan prior to WW2.

1. **FORT LONCIN:** The large fort was built on the western side of Liege. As described, it was destroyed when a 420mm round from a Big Bertha howitzer hit an ammunition bunker during the siege. More than 300 Belgians were killed instantly and today it is their resting place, a museum, and a memorial.
   N 50 40.422 E 5 29.592 (50.673700, 5.493200)

2. **FORT LANTIN:** Lantin was a small fort that was captured on August 15 after receiving some damage from the howitzers. The fort is in the shape of a triangle and had a turret with twin 150mm cannons, two 120mm cannon turrets, and various 57mm cannons for enfilading fire. After the war, it was not upgraded and was not used during WW2. Today, you can visit it to see its original status as in 1888.
   N 50 41.623 E 5 31.565 (50.693716, 5.526078)

3. **FORT LIERS:** This was a small fort that housed one turret with a single 210mm cannon, one with a 150mm cannon, and one with twin 120mm cannons, along with various 57mm cannons for local defense. In WW1, it was captured on August 14, after the Germans attacked it from the rear. It was not upgraded between the wars, but was used as storage for ammunition during WW2. Today, it is not accessible because a private company is using it for testing jet engines. I only include its coordinates for reference.
   N 50 41.922 E 5 35.036 (50.698700, 5.583931)

4. **FORT PONTISSE:** It is considered a large fort and had one turret with a 210mm cannon, one with twin 150mm cannons, and another twin turret with 120mm cannons. It also had four other turrets, each with 57mm cannons and 8 rapid-fire 57mm cannons. On August 12, it surrendered after several hits by the German 420mm howitzer. It was upgraded between the wars and captured during WW2 on May 18, 1940. Today, it is used as a refuge for bats and can be visited in the summer; if you want to!
N 50 41.529 E 5 38.368 (50.692148, 5.639474)

5. **FORT BARCHON:** This fort was one of the large ones, and is located on the eastern side of Liege. On August 5, it was unsuccessfully attacked by German infantry, but on August 8, it was attacked by the large howitzers. By the end of that day, the garrison surrendered when they were no longer able to breathe because of the dust and powder residue, because like all of the forts, there was no forced ventilation. The fort contained 2 turrets with 210mm cannons, 1 with twin 150mm cannons, 2 with twin 120mm cannons, 4 57mm cannon turrets, and 9 rapid-fire 57mm turrets. Of the 300-man garrison, 22 were killed. After the war, the Belgians upgraded the fort with new armament. The fort was attacked May 10-18, 1940 during WW2, before its surrender after damage by artillery, air attacks, and flamethrowers. Today, it is accessible for exploration and there is a war memorial to the Belgian defenders.
N 50 40.323 E 5 41.411 (50.672058, 5.690183)

6. **FORT EVEGNEE:** Evegnee was one of the "small" forts, built on the eastern flank of Liege. It contained a turret with a 210mm cannon, 1 with twin 15mm cannons, 2 with twin 120mm cannons, 4 turrets with 57mm cannons and 5 rapid-fire 57mm cannons. Being on the eastern edge of the battle, it saw intense action and was assaulted directly on August 10-11. The garrison surrendered at the end of August 11. After the war, it was upgraded and used for storing ammunition. During WW2, it was attacked from May 16 to 19, 1940, when it surrendered. Today, it is not accessible to the public and is used for military testing.
N 50 38.754 E 5 42.660 (50.645908, 5.710992)

7. **FORT FLERON:** It was one of the large forts for defending Liege, with turrets containing 210mm, 150mm, 120mm and 57mm cannons. After the Germans broke through to Liege, however, as with the other forts, they attacked Fleron from behind with siege cannons. The garrison fought until August 13, after being fired upon by the 420mm howitzer. It was upgraded prior to WW2 and during that war provided supporting fire from May 10-17, 1940. On May 17, after devastating air attacks, the commander received permission to evacuate and join the field army. Today, it is mostly buried under a public park that is accessible to walk around, and you can see the remains of the bunker and a plaque to the soldiers who fought and died there.
N 50 37.099 E 5 41.519 (50.618314, 5.691980)

8. **FORT CHAUDFONTAINE:** A "small" fort, it had a turret armed with a 210mm cannon, 1 with twin 150mm cannons, and 1 with twin 120mm and 57mm cannons. As did the other forts, it kept fighting until the siege cannons arrived. On

August 13, a shell caused a turret to explode, killing 58 soldiers, and shortly afterward, the garrison surrendered. The attack began on August 12. After the war, it was upgraded prior to WW2 when it was attacked by air and land, until the garrison surrendered on May 17, 1940. Today, it is an adventure park. Outside the main gate is a memorial to the soldiers and a small cemetery.
N 50 35.508 E 5 38.440 (50.591803, 5.640661)

9. **FORT EMBOURG:** Also a small fort, with the same armaments of 210mm, 150mm, 120 mm, and 57 mm cannons. It survived 48 hours of bombardment before surrendering. After the war, it was also upgraded. During WW2, it was attacked by ground, air, and artillery starting on May 13, 1940 when it was surrounded. On May 17, the garrison surrendered after most of the guns were put out of action. Today, it is a museum and has been preserved by volunteers for public visits.
N 50 34.967 E 5 37.077 (50.582781, 5.617955)

10. **FORT BONCELLES:** A large fort, Boncelles was armed with 2 turrets with 210mm cannons, 1 with twin 150mm cannons, 2 with twin 120 mm cannons, and 4 with 57mm cannons and 8-rapid fire 57mm cannons. After some severe bombardments, it was taken on August 14. After the war, it was upgraded. During WW2, it held out and did not surrender, forcing the Germans to take it by force, on May 16, 1940. Today, it is buried under a park, but there is a small cemetery, memorials, and a recently opened museum. The memorial is off the main street and you can venture around the corner for another entrance and path on the northeastern side of the fort.
N 50 34.680 E 5 31.890 (50.577999, 5.531503)

11. **11. FORT FLEMALLE:** Another large fort, it was armed with 2 turrets with 210mm cannons, 1 turret with twin 150mm cannons, 2 turrets with twin 120mm cannons, 4 with 57mm cannons and 11 57mm rapid-fire cannons. During the it was one of the last forts to come under bombardment and as such, it was the last to surrender, on August 16, only two hours after Hollogne. After the war, it was upgraded. During WW2 it was attacked by air by the Germans on May 15, 1940 destroying all its turrets and resulting in the surrender of the fort on May 16. Today it has memorials, a park and a museum.
N 50 36.494 E 5 27.989 (50.608233, 5.466481)

12. **FORT HOLLOGNE:** Another small fort, it was armed with a turret with a 210mm cannon, 1 with twin 150mm cannons, 1 with twin 120-mm cannons and 7 of the 57mm rapid-fire cannons. It came under bombardment on August 13, and surrendered on August 16. After the war, no upgrades were made and it was only used as storage for ammunition. Today, volunteers have bought it, cleaned it and made it into a museum with specific opening times.
N 50 38.778 E 5 27.962 (50.646306, 5.466028)

13. **CITADEL LIEGE:** In 1255, a fortress was built at this location overlooking Liege and the Meuse River. Ever since, the leaders of Liege have built and rebuilt defenses here. In 1650, the fort was transformed into a pentagon, destroyed by the

French, and rebuilt in 1684. During WW1, it was used as a command post and barracks, since the main defense was based on the circle of 12 forts. Today, the majority of the terrain has been made into a hospital. At the geo coordinates provided, however, there are 197 crosses as a memorial to the Belgians executed by the Germans during WW1 and WW2.
N 50 39.205 E 5 34.702 (50.653420, 5.578366)

14. **FORT DE LA CHARTREUSE:** This large fort was built after the Napoleonic War, which destroyed a monastery at the location. The Belgians used the site as a fort until 1891, when it was changed into a barracks. It didn't play a major role in either world wars, and was only used as a barracks and prison. Today, it is part of a city park and largely abandoned.
N 50 37.956 E 5 36.056 (50.632599, 5.600931)

15. **LIEGE ARMS MUSEUM, GRAND CURTIS:** Liege is famous for its weapons industry, and this museum has more than 13,000 specimens. It covers a wide range of history and is well worth a visit.
N 50 38.821 E 5 35.041 (50.647017, 5.584009)

16. **INTER-ALLIED MEMORIAL, LIEGE:** At this location is an Inter-Allied memorial dedicated to the Allied warriors of WW1. It is located in Liege because Liege was the first city to defend against the German invaders. The memorial is behind the Basilica of the Sacred Heart, and was refurbished in 2014 for the 100th Anniversary. A memorial service held there was attended by many heads of state and royalty.
N 50 37.169 E 5 34.138 (50.619477 5.568961)

FORTIFIED POSITION NAMUR: The following are 2 of the 9 forts the Belgians built around Namur. Only 2 are available to visit at this time; most are on private property. As stated previously, the Germans attacked Namur shortly after Liege and had success using the howitzers prior to an infantry assault. On August 16, with 107,000 soldiers, they surrounded the 37,000 Belgians defending Namur. The Germans began their artillery barrage of Namur on August 20, and it lasted until August 24, when the Belgians surrendered. In addition, the Germans had stopped a French relief force at Charleroi, leaving the Belgians in Namur with few choices. After the war, the Belgians upgraded most of the forts. The following are two forts that can be visited in Namur:

17. **FORT DE SAINT HERIBERT:** The fort was armed with 2 turrets with 210mm cannons, 1 with twin 150mm cannons, and 2 with twin 120mm cannons and four 57mm cannons. It came under intense bombardment starting on August 21, and it surrendered after heavy damage on August 24. After the war, it was upgraded. During WW2 it came under attack on May 15, 1940 from German infantry, aircraft, and artillery. The fort resisted until May 21, after inflicting more than 700 German casualties. Today, it is privately owned and is being refurbished by volunteers; it is only open on the 4th Sunday of the month between April and October. This geo coordinate is on the main road in front of the private road access gate.
N 50 24.814 E 4 49.541 (50.413570, 4.825685)

18. FORT DE EMINES: The fort was armed with 1 turret with a 210mm cannon, 1 with twin 150mm cannons, 2 with 120mm cannons, and 3 with 57mm cannons and 6 rapid-fire 57mm cannons. It began receiving fire from the Germans on August 23. After being hit by more than 2,000 shells in a little over 1 day, the Belgians surrendered on August 24. After the war, it was not upgraded, just repaired and designated as ammunition supply. During WW2 there was no major fighting for the fort.
N 50 30.433 E 4 51.278 (50.507213, 4.854632)

NATIONAL REDOUBT ANTWERP, SIEGE OF ANTWERP, SEPTEMBER 28-OCTOBER 10, 1914: As stated, Antwerp has always been a fortified city. After the Belgians gained their independence in 1830, they decided upon a defensive strategy for a "national redoubt" and Antwerp was chosen for it. After the Franco-Prussian war of 1870, the Belgians feverishly upgraded their system, including the forts of Namur and Liege, to deter further German or French aggression. In total, Redoubt Antwerp was built with 2 rings, starting in 1859, and with the major additions ending in 1914. Prior to 1859, forts were built during the 80 Years War, after Belgian independence in 1830, and in 1851, after the election of Napoleon III in France, when they built forts 1 through 7 around Antwerp.

In 1859, the redoubt construction began with a siege wall and 8 more forts. In 1870, the advent of German longer-range cannons, as used in the siege of Paris, caused the Belgians to realize their forts were too close to Antwerp. Therefore, they built Merksem, Zwijndrecht, and Kruibeke, and further reinforced the Scheldt forts. From 1878 to 1905, they built 11 more forts/redoubts to further guard the city and transit/rail lines.

Because of the advances in technology and the destructive power of weapons, however, the Belgians reviewed their fortress situation in 1900 and found them to be inadequate. As a result, under the Act of 1905 they began constructing 11 additional forts and 12 redoubts, completed in 1909. As with the other Liege and Namur forts, the Antwerp forts were built for long-range defense, inter-fort support, and close-in canal/self-defense. At the outbreak of WW1, the Belgians knew their forts would not withstand the German super howitzers, but they bravely stuck with their plan and staunch defense. After the Germans captured Liege, the Germans marched on to France, leaving behind a contingent to guard Antwerp. On September 22, the Germans returned to Antwerp to begin the siege in earnest after 2 major attempts by the Belgians to break out and attack the German flanks. On September 28, the Germans began using the heavy siege cannons directed by observation balloons. By September 30, they destroyed forts Walem, Sint Katelijne, Waver, and Koningshooikt. Next, Fort Lier was destroyed on October 3, Fort Kessel on October 4, and Fort Broechem on October 6. On October 9, the Belgians began evacuating Antwerp toward the neutral Netherlands and to the Ypres line via boats. Before they departed, however, they blew up forts Brasschaat, Kapellen, Lillo, Merksem, and Schoten. The Germans had won by firing nearly 600 shells from the 420mm cannons and more than 2,100 from the 305mm cannons. Among the 63,000 garrison troops and 87,300 field troops (Belgian and British), they suffered an unknown amount of casualties, 30,000 were captured, and 33,000 interned for the war in the Netherlands. The Germans attacked with 66,000 troops, and its casualty numbers also are unknown. They captured large amounts of equipment, including 2,500 cannons and swathes of ammunition.

After the war, everyone realized the forts would not withstand advances in technology. They still upgraded them for storage and infantry support and for longer-range cannons, and supplemented their defenses with tank ditches around Antwerp. During WW2, the Belgian Army decided to retreat from Antwerp to Westhoek because the Germans had broken through the Ardennes and defeated Fort Eben Emael. After defeating the Allies in 1940, the Germans occupied the forts for various reasons, one of them Fort Breendonk, which was used as a prison, a transit camp, and an execution site. Here are highlights of just some of the major forts to see; there are many more:

19. **ANTWERP, FORT BREENDONK:** The fort was built as part of Act 1905 and was a second-ring fort around Antwerp. During the siege of Antwerp; it fought against the Germans and was bombarded from October 1-9 until it surrendered. After the war, it was not really upgraded other than making it a little more useful and livable. During WW2, after the invasion of Belgium, the Belgians evacuated the fort prior to battle with the Germans over the fort. The Germans took it over and made it a Nazi prison camp where they transferred, imprisoned, tortured, and executed prisoners. In total, 3,500 prisoners transited the camp, 300 died there, and of those, 185 were executed. Today it has a museum and it is a national memorial.
    N 51 03.447 E 4 20.504 (51.057450, 4.341730)

20. **FORT LIEFKENSHOEK, ANTWERP:** Initially built in the 1580s, it was developed further during the 1800s. Today there is a museum with the history of the fort, which has been important since its construction in 1580 to World War II when it was part of the defensive ring around Antwerp. Walking around outside the fort is a great experience and they have an on-site cafe that serves the locally made beer "Bonapart."
    N 51 17.667 E 4 17.133 (51.294444, 4.285556)

21. **FORT LIEZELE:** It was finished in 1914, in time for the start of WW1, and participated in the defense of Antwerp in October, 1914. During WW2, it had a rather insignificant role because the Belgians evacuated Antwerp before the Germans arrived. Today, it is the best-preserved fort and has an excellent museum and collection. However, as with many of the forts, you must check the opening times because they are generally operated by local volunteers for one Saturday a month.
    N 51 04.061 E 4 17.135 (51.067688, 4.285577)

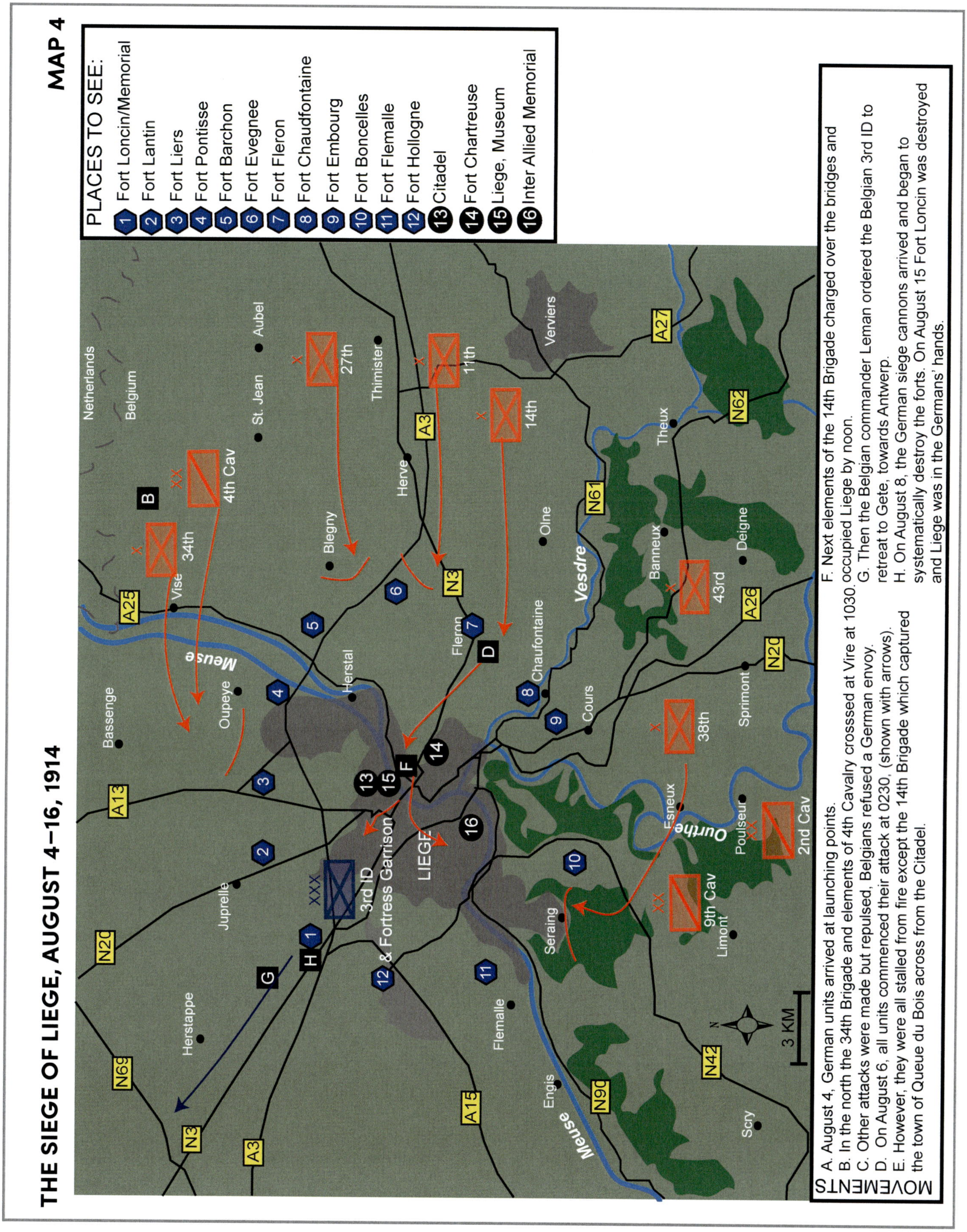
THE SIEGE OF LIEGE, AUGUST 4–16, 1914
MAP 4
PLACES TO SEE:
1 Fort Loncin/Memorial
2 Fort Lantin
3 Fort Liers
4 Fort Pontisse
5 Fort Barchon
6 Fort Evegnee
7 Fort Fleron
8 Fort Chaudfontaine
9 Fort Embourg
10 Fort Boncelles
11 Fort Flemalle
12 Fort Hollogne
13 Citadel
14 Fort Chartreuse
15 Liege, Museum
16 Inter Allied Memorial
Netherlands
Belgium
Aubel
St. Jean
Thimister
Verviers
Bassenge
Vise
Oupeye
Herstal
Blegny
Herve
Theux
Fleron
Olne
Chaufontaine
Vesdre
Banneux
Deigne
Meuse
Cours
Sprimont
Esneux
Ourthe
Poulseur
Juprelle
LIEGE
Seraing
Limont
Herstappe
Flemalle
Engis
Scry
4th Cav
34th
27th
11th
14th
43rd
38th
2nd Cav
9th Cav
3rd ID
& Fortress Garrison
A25
A13
N20
N69
N3
A3
A15
N90
N42
N61
A27
N62
A26
B
D
F
G
H
3 KM
N
MOVEMENTS
A. August 4, German units arrived at launching points.
B. In the north the 34th Brigade and elements of 4th Cavalry crosssed at Vire at 1030.
C. Other attacks were made but repulsed, Belgians refused a German envoy.
D. On August 6, all units commenced their attack at 0230, (shown with arrows).
E. However, they were all stalled from fire except the 14th Brigade which captured the town of Queue du Bois across from the Citadel.
F. Next elements of the 14th Brigade charged over the bridges and occupied Liege by noon.
G. Then the Belgian commander Leman ordered the Belgian 3rd ID to retreat to Gete, towards Antwerp.
H. On August 8, the German siege cannons arrived and began to systematically destroy the forts. On August 15 Fort Loncin was destroyed and Liege was in the Germans' hands.

Photo left is a memorial to the fallen Belgian warriors that defended Liege at the start of WW1. The memorial is located outside the remains of Fort Loncin, the last fort to surrender during the Battle of Liege. Photo above right is of the fort entrance whch is now part of a museum and memorial. Below left is the Liege Arms Museum, which provides a wide array of weapons from throughout history. Below photos show just some of the many weapons and armor on display at the museum.

BATTLE OF MONS

August 23-24, 1914

"But if our hopes are betrayed, if we are forced to resist the invasion of our soil and to defend our threatened homes, this duty, however hard it may be, will find us armed and resolved upon the greatest sacrifices." -King Albert II

SUMMARY: After the capture of Liege and Namur, the German war machine advanced on Mons, Belgium, fighting through the courageous defense of the small and outgunned Belgian forces. On August 23, the British Expeditionary Force (BEF) swiftly assembled a defense there and fought to hold back the German war machine. Despite the British courage, they were forced to withdraw to reform a defense near Paris, after a German breakthrough on their flank.

As the Germans fought their way through the Belgian forces, the BEF landed more troops in France and organized a defensive line near Mons. Their mission had been pre-planned in case of war and was coordinated with the remaining Belgian forces to hold the Allied left flank from the coast to the French fortress town of Maubeuge, where the French 5th Army held the line. By August 22, 4 British infantry divisions and one cavalry division were in France, taking up positions right in the path of the strongest part of the German offensive, General Heinrich Rudolph von Kluck's 1st Army.

The German Schlieffen Plan called for a strong right flank to break through Liege to capture it, Brussels, and Namur, and move on to Paris. Between Namur and Paris lay Mons, where the British prepared their defensive positions along the Mons-Conde Canal. The British I Corps was on the right flank under the command of General Douglas Haig, the II Corps was in the center under the command of General Horace Smith-Dorrien, and on their left was the 19th Infantry Brigade. Beyond the brigade, to the north were the remaining Belgian forces defending the line up to the channel. From a British perspective facing east, the German 1st Army advanced on Mons with the German II Army Corps on the right at Conde, IV Army Corps in the center at Pommereuil, and III Army Corps on the left at Tetre and along the canal bend. Finally, on the far right, the IX Corps pressed on the southern side of Mons, from Obourg.

On August 23, with the Allied defensive positions partially prepared, the Germans began an artillery bombardment. At 0530, the overall British Commander Sir John Denton Pinkstone French (Yes, the British commander's last name was French!) met with Haig and Smith-Dorrien at his forward HQ. He ordered the canal line to be reinforced and the bridges prepared for demolition. Between 0600 and 0700, the German cavalry patrols skirmished with the British near Nimy and Pommereuil. The British near Obourg exchanged shots with the Germans. At 0900, the Germans began shelling the British canal positions and the IX Corps engaged heavily with the 4th Middlesex Regiment at Obourg. The British rifle and machine-gun fire inflicted heavy casualties on the Germans. By 1100, the German III Corps was pressing the British line at Jemappes, but the Germans were also taking heavy casualties. Near there, the Germans pushed the 1st Royal West Kents Regiment back over the canal from Tetre. By noon, the German front had widened to 11 km. Shortly after, the Germans crossed at Obourg and reached the railway line, leaving the Middlesex regiment and the 2nd Royal Irish Regiment in dangerous positions.

At 1400, German cannons targeted the British positions at Nimy, including the 4th Royal Fusiliers Regiment, which forced them to withdraw from the town, allowing the Germans to cross the canal. The Fusiliers' 2 machine guns at Nimy were under the command of Lt. Maurice Dease, who earned one of the two Victoria Crosses awarded for the battle.

As it raged and all of Dease's men were either wounded or killed, he took over the Vickers machine gun. As he fired and held his position against the attacking Germans, he was wounded 5 times, but did not relinquish until he was taken to an aid station where he subsequently died. After his departure, Pvt. Sidney Godley jumped into the fray and fired the same machine gun to help cover the battalion's withdrawal, and he was also wounded in the process. When he could no longer stop the German advance, he dismantled the MG and threw it in the canal before being captured. Godley was awarded the other Victoria Cross for his heroism. In spite of it, by 1500 the British right flank was in dismal shape. The 3rd Division was pulled off Hill 93 southeast of Mons and replaced by 2 brigades. The French cavalry to their right, however, fell back and on the left flank the 1st Royal Scots Regiment was ordered to retreat from Jemappes. As a result, the Germans were able to cross the canal at Jemappes.

In the late afternoon on the right flank, the Middlesex and 2nd Royal Irish Regiments withdrew from Obourg back to Bois la Haut because the Germans had begun to outflank them. In the center, the Germans renewed an attack against the 8th Brigade at 1900, and forced them to withdraw to Nouvelles. Around dusk, the Germans called a ceasefire to consolidate forces. Close to 2100, British BEF Commander French ordered the II Corps to hold its position. Then early morning of August 24, Commander French was informed that Tournai, on his southern/right flank, had fallen and that the French forces were pulling back. The British had little choice; they had to withdraw before being surrounded and form a new defensive line near Cambrai.

OUTCOME: The BEF was successful in slowing the German Army and giving the French forces additional time to retreat and reorganize. Also to note, we usually talk about German atrocities in WW2, but in fact, the Germans retaliated against many civilians and POWs in Belgium in WW1 also, executing citizens and burning homes. However, there was no Nazi ideology during WW1 to justify the murders; it just came out of frustration with the Belgians for putting up such stiff resistance.

WHO? NUMBERS AND LEADERS? The British Expeditionary Force under command of Sir John French comprised 80,000 men contained in: 1 Cavalry Division, the 5th Cavalry Brigade, 19th Infantry Brigade, I Corps with the 1st and 2nd Infantry

Divisions, and II Corps with the 3rd and 5th IDs. Each division had more than 18,000 men, made up of 3 brigades of 4 battalions. The I Corps was under the command of General Haig and the II Corps under the command of General Smith-Dorrien. The Germans attacked with General von Kluck's 1st Army, which contained 4 active army corps: II, III, IV, and IX, plus 3 reserve corps: III, IV, and IX. The Germans had a total of 150,000 men. In all, the British suffered approximately 1,600 casualties, the Germans about 4,000.

WHEN? August 23-24, 1914

WHY AT MONS? The Germans put their Schlieffen Plan into action, which required their right flank's assault to be extremely strong; the strongest part of the attack bypassed Brussels and went through Mons.

TECHNOLOGY/TACTICS: The battle of Mons presented the Allies with an extremely complicated defensive position. First, the Belgian forts did not last as long as they should have and 2nd, the French on the British right flank began to fall after the British were rather successful in holding back the Germans. Both sides had about the same technologies of bolt-action rifles and machine guns. Because the Germans were on the offensive, they did not necessarily have their machine guns placed for covering fire, but rather their infantry advanced with only some artillery barrages, followed by an advance on foot. The foot soldiers were met by British and French machine guns, which inflicted heavy losses on the Germans. Eventually the Germans' sheer weight of men and supplies overpowered the Allies and forced them to withdraw.

PLACES TO SEE:

1. **MONS MILITARY HISTORY MUSEUM:** Recently refurbished, the Mons military museum is exceptional. The displays include information and maps on the battles that took place in the area, including: the 80 Years War, the Spanish Succession, the Austrian Succession, and the Napoleonic Era battles. The majority of the museum, not surprisingly, is dedicated to the World Wars.
   N50 27.004 E 003 57.390 (50.450067 3.9565)

2. **MONS, 1ST AND LAST SHOTS FIRED MEMORIALS:** Outside of Mons are memorials to the 1st British shots fired during WW1 (their first on the continent since Waterloo) and a memorial across the street to the last British shot fired during the war.
   N 50 30.307 E 3 59.896 (50.505115, 3.998262)

3. **BERGEN MONS COMMONWEALTH AND ALLIED CEMETERY:** In the middle of the old civilian cemetery are the British Commonwealth Memorial and Cemetery and a Belgian soldiers' cemetery with memorials. This combined cemetery contains the graves of 383 Commonwealth soldiers. The geo coordinate is the parking lot. Upon entering the cemetery, you need to walk up to the left and onto the diagonal path to the rear of the cemetery to see the military cemetery and memorials.
   N50 27.530 E3 58.537 (50.458833 3.975617)

4. **ST. SYMPHORIEN COMMONWEALTH CEMETERY:** This cemetery commemorates 229 Commonwealth and 284 German servicemen who died during WW1. It contains the graves of the 1st and last Commonwealth soldiers believed to have died in the Great War. Those soldiers were Pvt. John Parr, killed during a patrol before the Battle of Mons started, and George Ellison and George Price, killed on the last day of the war, November 11, 1918.
N 50 25.950 E 4 00.633 (50.432500, 4.010556)

5. **MONS, CANADIAN LIBERATION MEMORIAL PLAQUE AND BRITISH MEMORIAL FOR THE BATTLE OF MONS:** On the entrance wall of the city hall there are 2 large plaques. One is for the British Regiments that fought in the 1914 Battle of Mons, the 2nd is dedicated to the Canadians who liberated Mons in November, 1918.
N 50 27.291 E 3 57.134 (50.454847, 3.952239

6. **MONS, US LIBERATION MEMORIAL FOR WORLD WAR II:** At this location is a memorial to the US 1st Infantry Division that liberated Mons during World War II, on September 6, 1944. It lists the names of the fallen who died from July 25 to the liberation of Mons.
N 50 27.034 E 3 57.459 (50.450560, 3.957643)

7. **MONS, MACHINE GUN MEMORIAL:** On the canal and under a bridge there is a memorial commemorating the heroic action of Lt. Dease, who, despite several wounds, manned a machine gun to hold back 2 German Battalions. After he was taken away (he later died of his wounds), Pvt. Godley, also wounded, took over the gun, eventually dismantling it and throwing it in the river before being captured. Both men were awarded the Victoria Cross.
N 50 28.518 E 3 56.724 (50.475302, 3.945404)

8. **MONS, CELTIC CROSS AND CROSSROADS:** During the battle, the Germans attempted to cut off the British retreat. At this location, elements of the Royal Irish Regiment, with 40 men and a repaired machine gun, held off the Germans trying to encircle their flank.
N 50 27.130 E 3 58.807 (50.452174, 3.980114)

THE BATTLE OF MONS, AUGUST 23–24, 1914

MAP 5

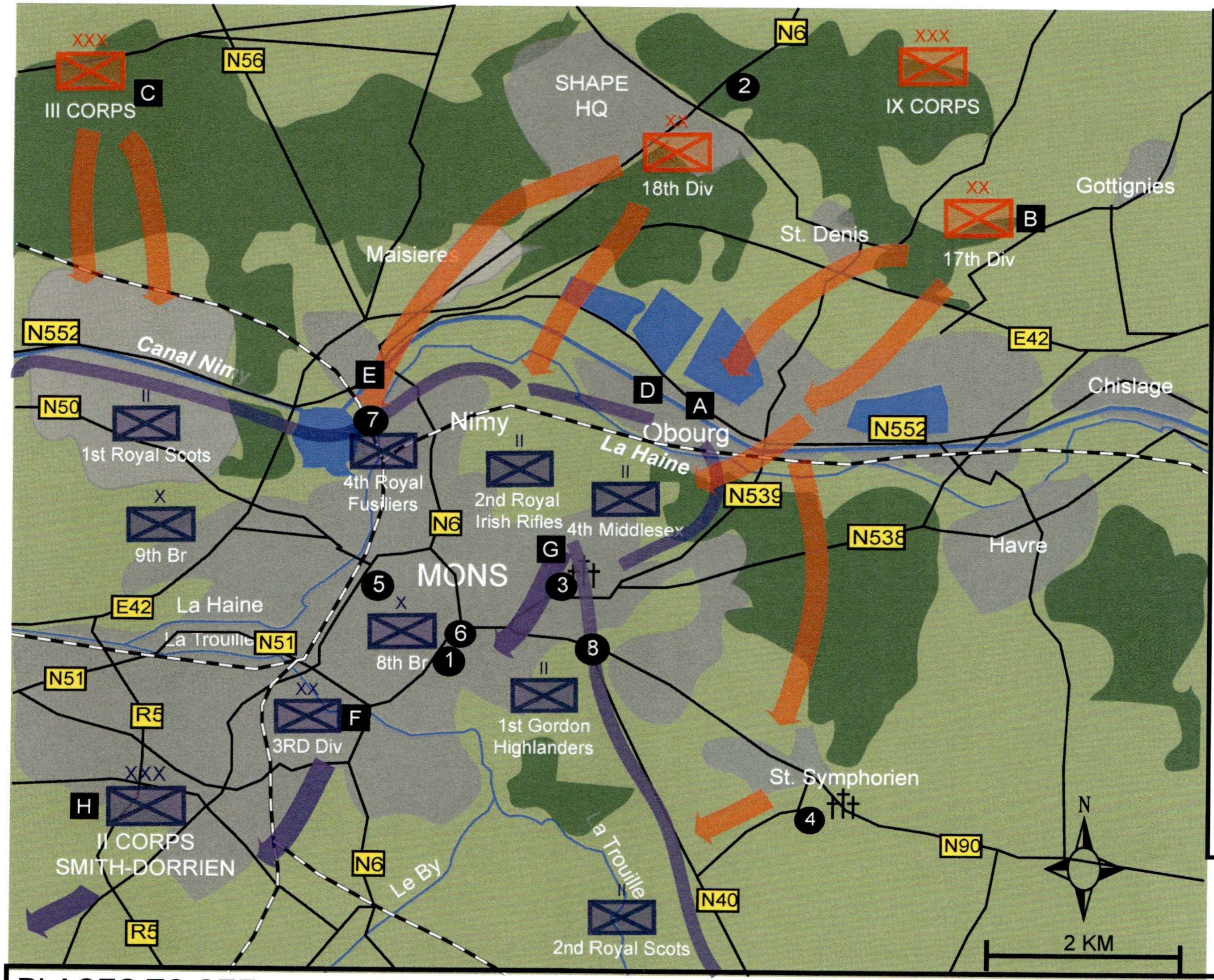

MOVEMENTS:

A. On August 23, the Germans and British exchanged shots near Obourg.
B. After a German artillery barrage at 9am, the German 17th Division began attacking the 4th Middlesex at Obourg.
C. At 11am, to the west the German III Corps began to press the British line at Jemappes.
D. At noon, the Germans crossed the canal at Obourg.
E. By 1400, after being pounded by the German artillery, the 4th Royal Fusiliers were being forced to withdraw from the canal.
F. At 1500 the majority of the 3rd Division was ordered to pull back, while two brigades covered the retreat. And the 1st Royal Scots of the 9th Brigade were ordered back.
G. At the same time the 2nd Irish and the 4th Middlesex retreated from Obourg to avoid being encircled.
H. After midnight, the French on the right flank begin to retreat. The British with little choice had to order a full retreat.

PLACES TO SEE:

1. Mons Military Museum
2. Mons, First & Last Shot Mem.
3. Mons, Commonwealth Cem.
4. St. Symphorien Cem.
5. Mons Town Hall Plaques
6. Mons, WWII Liberation Mem.
7. Machine Gun, VC Action Memorial
8. Celtic Cross, Crossroads Memorial

Photo left is a memorial to the "first shots" of the War for the Commonwealth. Across the street is memorial and plaque to the last Commonwealth shots fired during the war. These are located just east of the beautiful downtown of Mons as seen in the below two center photos. The photo to the right is of the town square. The townhall in the center holds several memorial plaques on the entrance.

The photo left is of the cemetery of St. Symphorien which uniquely has the graves of German and Commonwealth soldiers. The cemetery was created by the Germans shortly after the Battle of Mons. Today it holds the graves of the Victoria Cross recipient Lt. Dease who bravely defended the canal crossing with a machine gun until he was mortally wounded. It also has the first and last soldiers that died during the war for the Commonwealth.

FIRST BATTLE OF THE MARNE

September 5-12, 1914

"My center is giving way, my right is retreating, situation excellent, I am attacking."–Ferdinand Foch

---

SUMMARY: After a grueling month of German victories, the French Army and British Expeditionary Force stopped the Germans at the Marne River only kilometers to Paris. A victory known as the "Miracle on the Marne." The "miracle" forced the Germans back on their heels and a retreat to a defensive posture and the instigation of trench warfare.

Since the war's onset, the Germans had been defeating the Allies, who were in a disorganized retreat to defend Paris. On September 4, the city was in a panic, the Germans 1st and 2nd Armies were within 64 km of Paris It appeared to be a repeat of embarrassing Franco-Prussian War of 1870. However, instead of driving on into Paris, the center of power to end the war, the Germans fatefully turned to envelop and destroy the Allied armies.

The maneuver exposed the German right flank to French forces in and around Paris. French Commander-in-Chief Joseph Joffre recognized the opportunity and organized a counterattack. On September 5, on a 300-km front, the French regrouped and attacked along the Marne with 1,071,000 soldiers against Germany's 1,485,000. At that moment, facing south, the German 1st Army commander recognized the counterattack from Paris and wheeled right to meet the initial threat from the French 6th Army on the Germans' right flank. In doing so, the German commander created a 48-km gap between the 1st and 2nd Armies. The Allies realized the opportunity to encircle the German forces and attacked the gap.

By September 9, the Allies were close to encircling some of Germany's forces. On hearing the news, German Commander Helmuth von Moltke nephew of the 19th-century German commander of the same name (he captured Paris in 1870), had a nervous breakdown and his subordinates called an all-out retreat to the Aisne River, 64 km north. The Germans dug in at the Aisne River and trench warfare began. After failing to break through, Moltke was quoted as saying to the Kaiser, "Your Majesty, we have lost the war."

OUTCOME: The German Schlieffen Plan was stopped and Paris was saved. The Germans and Allies "raced to the sea" in an attempt to outflank each other. After neither sider were able to out flank the other, a stalemate was created and trench warfare initiated. The casualties were immense; mankind had applied his industrial wisdom to the destruction of fellow man.

WHO? The Allied force was helmed by the French Commander-in-Chief Joffre, with a million men at his disposal. English Commander French controlled 70,000 men. The Allied casualties combined numbered 263,000. In comparison, the German force of 1,485,000 men commanded by Moltke and Von Kluck suffered 220,000 casualties. (Note: The Germans would have had 3 more infantry corps in the battle, but they had to stem a Russian breakthrough in Poland/Prussia. With those additional troops on the Eastern Front, the Germans quickly turned their dire situation there into an astounding success at the Battle of Tannenberg.)

WHEN? September 5-12, 1914

TECHNOLOGY/TACTICS: The Germans used the Schlieffen Plan to great success until they changed their tactics to stop the advance on Paris. The shift allowed the French to reorganize and counterattack. The French even used Paris cabs to send 6,000 reinforcement troops to meet the German threat. WW1 saw the use of new weapons of warfare, including machine guns, breach-loading cannons, shrapnel rounds, poison gas, smokeless ammunition, barbed wire, airplanes, zeppelins, U-boats, battleships, and tanks. It was murderous warfare on a level never seen before.

PLACES TO SEE:

1. **MEMORIAL DES BATAILLES DE LA MARNE, DORMANS:** This is a very beautiful chapel on the south side of the Marne, overlooking the site where the Germans and French clashed. It is located at what had been the French force's right flank.
   N 49 04.302 E 3 38.764 (49.0717, 3.646067)

2. **NATIONAL MONUMENT OF THE MARNE VICTORY, MONDEMENT-MONTGIVOUX:** Located at the Battle of the Marshes of Saint-Gond, a decisive point of the Battle of the Marne, it is a 33 meter-high monolithic monument with information panels and a small museum nearby. It was at the center of the French defensive line.
   N 48 47.193 E 3 46.511 (48.78655, 3.775183)

3. **MARNE MUSEUM AND AMERICAN MEMORIAL TO FRENCH COMBATANTS OF THE MARNE, "HOPELESS VICTORY":** At this location is an American-built memorial at the site where the French stopped the Germans. It is just outside Meaux, to the east of Paris. The Meaux Museum next to the monument opened in 2012. It is very well done, with open displays and equipment you won't see elsewhere. The museum also provides information about the Franco-Prussian War, which shifted Europe's power bases and alliances, and consequently set up the continent for WW1.
   Marne Museum: N 48 58.248 E 2 54.176 (48.9708, 2.902933) The American monument is next to the museum; just park at the museum and walk over.

4. **BRITISH MONUMENT AND MEMORIAL TO THE DEFENSE OF THE MARNE, CONDETZ:** The British Expeditionary Force and the French pulled back to hold the defensive line. The town of Condetz contains a monument to the British forces who fought and died there.
   N 48 56.586 E 003 7.454 (48.9431, 3.124233)

1ST BATTLE OF THE MARNE, SEPTEMBER 5–12, 1914

MAP 6

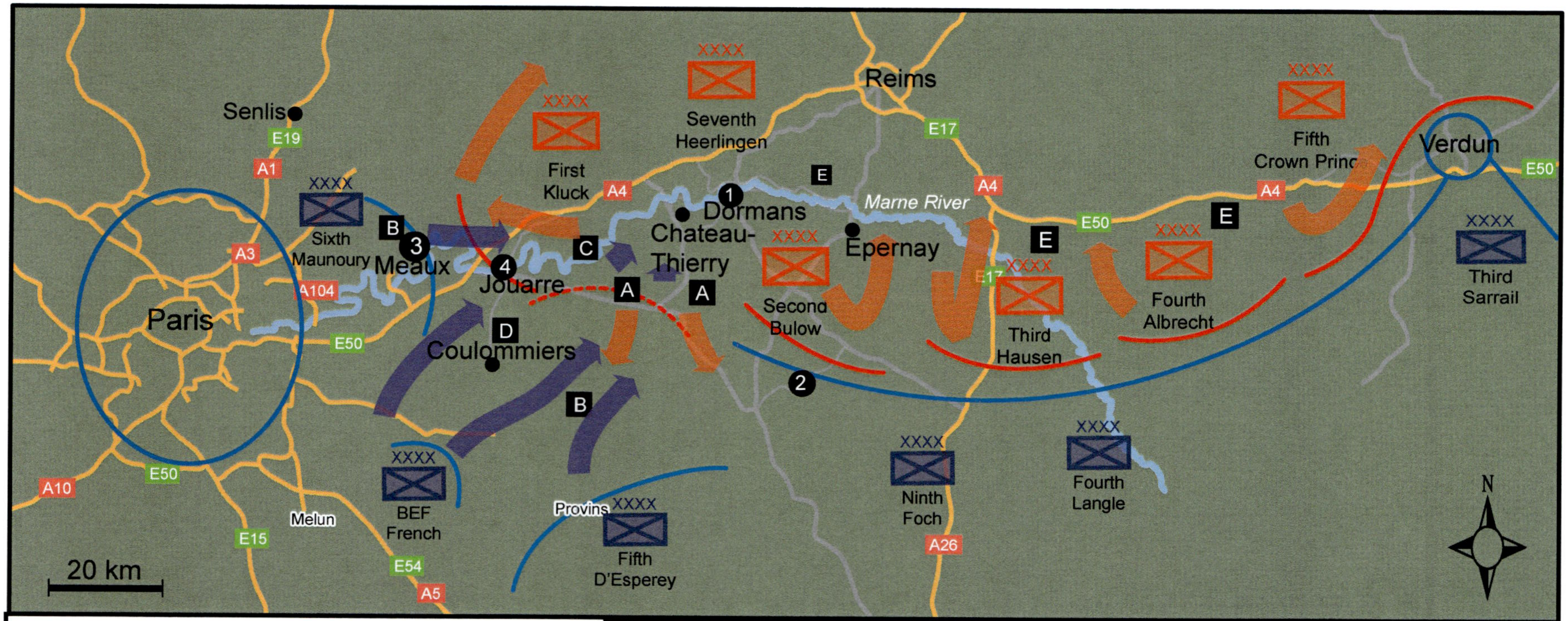

Places to See:

1. Memorial des Batailles de la Marne
2. National Monument of the Marne Victory
3. Marne Museum and Memorial to French Combatants of the Marne, "Hopeless Victory"
4. British Monument and Memorial to Marne Defense

MOVEMENTS

A. By September 5, the Germans began advancing on Paris but instead of staying on target to envelope Paris, they turned south to envelope the French Armies.

B. In response on September 5, the French and British counterattacked, and hit the German First Army in the right flank.

C. German First Army wheeled right to meet the threat but the move exposed a 60 KM gap between the First and Second Armies.

D. French and BEF attacked through the gap and threatened encircling the Germans.

E. The German Command ordered a retreat to avoid a total loss. The Germans set up their new defensive line on the Aisne River.

Upper left, Dormans, a beautiful chapel was built in memory of the sacrifices that took place during WW I. It overlooks the fields where the Germans and French clashed just south of the Marne River. Photo right, is the National Monument to the Marne victory. It is 33 meters high and overlooks the battlefield that was the French defensive center in September 1914.

Chateau Thierry, the powerful American memorial overlooks the Marne River at the location the American Forces, 3rd ID "Rock on the Marne" and 2nd ID pushed the Germans back in June 1918. Photo right is the "Hopeless Victory", memorial The statue was donated to the French by America in memorial to their defense against the Germans in 1914.

Right is an example of a beautiful and now peaceful fields around the Marne River This field is located near the Marne Victory monument. And of course in another field, some of the hefty French cows.

FLANDERS FIELDS

"In Flanders fields the poppies grow between the crosses, row on row..."-John McCrae

---

SUMMARY: Flanders Fields is a name given to the Flemish area of Belgium, which lies to the north of Brussels and runs along the coast. The area became famous with the help of the poem, "In Flanders Fields," by Canadian Medical Officer Lt. Col. John McCrae. The poem became one of the most popular poems describing the terrible, haunting war. There was a total of five battles in this small area over a four year period. I have provided a short summary of each of them and a corresponding map:

1ST BATTLE OF YPRES, OCTOBER 19-NOV 22, 1914: After the Germans' failed attack at the Marne near Paris, the 2 sides began a "race to the sea," where each tried unsuccessfully to outflank the other until they hit the northern coast of Belgium. During the flanking melee, the Germans captured Antwerp, and the British and Belgians retreated to the historic town of Ypres, which was at a critical road junction famous for its cloth production. There, the Allies formed a defensive line all the way to the channel.

On October 19, German Commander General Erich von Falkenhayn attacked with 2 armies. The objective was to overrun the Belgian sector in the north to capture Dunkirk and then Calais on the coast, cutting the British supply line. The Germans made gains initially, but on October 27 the Belgians opened floodgates and flooded the fields from the coast to north of Ypres, stopping the German offensive. Next, the Germans tried attacking Ypres with 1 army and 8 divisions against the British Expeditionary Force, which had 8 infantry divisions and 3 cavalry divisions. The British held against the onslaught, but the Germans managed to gain the high ground of Messines Ridge, which forms a dominating S-shaped ridge around the east and south of Ypres. Finally, on November 11 the Germans attacked with 2 divisions north of Menin Road, an area called "Nuns Wood." They managed to break through, but were hesitant and did not exploit their gains.

Soon after, the winter weather set in, making offensive operations all but impossible because of the mud and the terrible conditions. With little movement, both sides built up their defenses and shelters for trench warfare. The Germans built strong concrete positions that were more sanitary compared with the Allies, who believed their situation would only be temporary. The High Command refused to establish better living conditions, saying they wanted to "stay on the offensive."

During this battle, the British had 163,000 soldiers in the field and suffered 58,155 casualties (7,960 KIA) in the span of less than a month. Many believed the "professional British army" was lost in the first battle of Ypres. The French suffered more than 50,000 casualties and the Belgians over 20,000. The Germans also suffered greatly, with more than 134,315 casualties, 25,000 of them eager young patriotic university students. The Germans came to refer to the first battle of Ypres as "Kindermord bei Ypern"(massacre of the innocents at Ypres). Both sides were shocked by the new type of warfare and the amount of suffering it brought. The industrial revolution had truly modernized weapons and warfare for maximum effect.

2ND BATTLE OF YPRES, APRIL 22-MAY 5, 1915: After a failed Allied offensive in the Champagne region, there was a lull in major engagements until April 22, when the Germans used chemical weapons for the first time on the western front (they used it on the Russians in January 1917). On April 22, at approximately 1700, the Germans opened up with a large salvo of traditional artillery. Then they opened more than 5,700 canisters (160 tons worth) of chlorine gas on a 6.5-km front near the town of Gravenstafel, northeast of Ypres.

Chemical weapon deployment was still in its infancy and relied on the wind to disperse the gas. When the wind was right, the Germans opened the large canisters resembling industrial milk jugs. The greenish-yellow clouds crept along the front and sank into the French and Algerian trenches. It was a total surprise and asphyxiated more than 6,000 men in just 10 minutes. An additional 2,000 were blinded and taken as POWs.

The result left a 6.5 km-gap in the Allied lines, which the Germans were slow to exploit because they not only were surprised by the gassing's success, they were too frightened to advance while it could still be in the air. The lines were quickly and gallantly filled by the Canadians, who urinated on cloth and wrapped it around their faces to block the effects of the gas. On April 24, the Germans fired artillery and opened gas canisters again, but this time on the Canadians in the area of St. Julien Village, now called "Kitchener's Wood." The Canadians fought valiantly, earning high regard, but they suffered tremendously in the German onslaught and the disastrously effective chemical attacks. In the end, the Allies had to withdraw closer to Ypres and the Germans gained additional high ground to the north, creating a half-circle around Ypres.

In the battle, the Allies, who had 8 divisions present, approximately 130,000 soldiers, suffered more than 65,000 casualties total. Of that total, the Canadians more than half of their 10,000 soldiers committed. The British counted 50,000 casualties and the French more than 10,000. The Germans and their 7 divisions had more than 35,000 casualties. The battle was infamous for the first chemical attacks on the Western Front, but also famous for the poem "In Flanders Fields" written by Canadian Medical Officer Lt. Col. John McCrae:

"In Flanders Fields"

In Flanders fields the poppies grow
Between the crosses, row on row,
That mark our place; and in the sky
The larks, still bravely singing, fly
Scarce heard amid the guns below.

We are the Dead. Short days ago
We lived, felt dawn, saw sunset glow,
Loved, and were loved, and now we lie
In Flanders fields.

Take up our quarrel with the foe:
To you from failing hands we throw
The torch; be yours to hold it high.
If ye break faith with us who die
We shall not sleep, though poppies grow
In Flanders fields.

3RD BATTLE OF YPRES, MESSINES JUNE 7-14, 1917: Considered a masterpiece in planning, the Messines Ridge attack was a much-needed success for the Allied Forces. The ridge, which lies 10 km south of Ypres, had been in German hands since 1914, giving them complete observation of the British lines. Before the third battle started, since early 1915, the British forces (plus New Zealanders, Australians, and Irish) had dug tunnels and the Germans dug counter-tunnels. Famous battles ensued over slight ridges such as Hill 60 and Hill 62, where thousands of men fought and died above and below ground.

During the 3rd Battle of Ypres, British Commander Lt. Gen. Herbert Charles Onslow Plumer had great success because of his excellent planning, which included communication of objectives to the lowest levels, thorough air reconnaissance of the German locations and supply routes, accumulation of supplies (ammo, food, and water), and preparation of 19 large mines under the German positions. On June 7 at 0310, the Allied demolition teams triggered their plungers and 19 enormous explosions erupted along 16 kms of tunnel beneath the German lines (from Hill 60 in the north to Trench 122 in the south). Tons of debris and 10,000 dead Germans were thrown into the air in what was the biggest man-made explosion ever seen then, and later until the nuclear bomb. Next, an artillery barrage of 2,000 guns opened up on the front line to kill any survivors. Shortly afterwards, Plumer used a "creeping barrage" for the Allied soldiers to advance, and they quickly occupied Messines Ridge.

The Germans tried sending in reinforcements, but the Allies held the high ground and easily called in artillery on the advancing Germans. The Allies then attacked to capture their last objective 2 km east of the ridge. After a week of additional fighting, the Germans withdrew and the British had their best success of the war. The Allies, with 12 divisions under Plumer, and the Germans each suffered approximately 25,000 casualties.

3rd Battle of Ypres, Passchendaele, July 31-November 10, 1917: Passchendaele ranks with Verdun and the Somme as one of the most horrendous battles in history because of the fighting conditions, the extreme casualties, and the fact that strategically it was totally unnecessary. After the success of the Messines Ridge, the British were slow to continue the attack; in fact, they stalled for 6 weeks, allowing ample time for the Germans to fill in the gaps and strengthen their positions. The still-confident British Commander Douglas Haig was obsessed with capturing Belgium's channel ports to seize the U-boat safe havens and force the Germans from the country. Instead of keeping Plumer, who was highly effective at Messines Ridge, Haig removed him from command.

The 2nd phase of the 3rd Battle of Ypres started with a massive 10-day artillery barrage in which more than 4.25 million rounds were fired. As with other barrages, the idea was to blow up the German defenses and barbed wire. Instead, they combined with rainfall to make the already low sea-level battlefield into dirt porridge. Despite the conditions at 0350 on July 31, the British and French attacked across an 18-km front with 500,000 men, 136 tanks, and 500 aircraft. They were up against the German 4th and 5th Armies. The attack quickly slowed because of the landscape of thick mud and craters. On August 16, Haig halted the attack to re-align his front. At that point, no major breakthrough seemed possible, but Haig ignored the facts, still confident he could break the German line.

On September 20, Plumer was reinstated to renew the attack and he had some initial success with concentrated artillery bombardments, coupled with rolling barrages for the infantry. But, the German pillboxes with interlocking fire and their use of mustard gas took a heavy toll on the Australia and New Zealand ANZACs and the Canadians leading the way. No matter, the Allies sent 2 more Canadian divisions into the breech, and they finally captured Passchendaele on November 6. The casualties sustained for the ground taken is astonishing; in 16 days of fighting, it cost the Canadians alone 15,654 casualties for 800 meters, from Crest Farm to the center of Passchendaele.

After more than 3 months of fighting, the Allies effort only created a 12 km bulge into the German lines. Worse yet, as a result of the bulge, the Germans could fire from three sides, and within 6 months it was evacuated by the Allies. The losses were astounding; the British suffered more than 450,000 casualties, the Germans more than 270,000. As a result, the British command broke up in a lot of finger-pointing and questions about why they continued the attack after having had no success in August, and after everyone was aware that with the arrival of the German reserves and the rainfall, there was no chance of a major breakthrough.

Poem by Francis Ledwidge, 1917 Irish poet and soldier who fought and died during the battle:

"Soliloquy"

When I was young I had a care
Lest I should cheat me of my share
Of that which makes it sweet to strive
For life, and dying still survive,
A name in sunshine written higher
Than lark or poet dare aspire.

But I grew weary doing well.
Besides, 'twas sweeter in that hell,
Down with the loud banditti people
Who robbed the orchards, climbed the steeple
For jackdaws' eyes and made the cock
Crow ere 'twas daylight on the clock.
I was so very bad the neighbours
Spoke of me at their daily labours.

And now I'm drinking wine in France,
The helpless child of circumstance.
To-morrow will be loud with war,
How will I be accounted for?

It is too late now to retrieve
A fallen dream, too late to grieve
A name unmade, but not too late
To thank the gods for what is great;
A keen-edged sword, a soldier's heart,
Is greater than a poet's art.
And greater than a poet's fame
A little grave that has no name.

4TH BATTLE OF YPRES (BATTLE OF ESTAIRES/LYS), APRIL 9-19, 1918: The 4th Battle of Ypres was a result of the German Spring Offensive, called "Georgette." The German objective was to capture Ypres and push the British into the channel. The Germans offensive gained ground but they outpaced their logistics and artillery. Furthermore, some American units arrived to help stop any breakthrough.

5TH BATTLE OF YPRES, SEPT 28-OCT 2, 1918: The 5th Battle of Ypres took place during the 100 Day Offensive, the major Allied counteroffensive, and had great success against the demoralized Germans. The battle was led by Belgian King Albert I. Under his command, the Allied divisions gained 29 km and recaptured most of what was lost during the German Spring Offensive. They had to stop their advance, however, because of a supply shortages.

WHY IN FLANDERS? After the Germans were stopped outside Paris, the French together with the British, and the Germans tried to outflank each other all the way up to the channel. As a result, the two sides faced each other in Flanders, around the mining town of Lille and the major railroad junction of Ypres.

PLACES TO SEE: (ORGANIZED NORTH TO SOUTH)

NORTH:

1. **LANGEMARK, GERMAN CEMETERY:** This was the location of the German front lines, part of the southern thrust and gas attacks of the 2nd Battle of Ypres. After World War II, Belgium forced the Germans to consolidate more than 678 cemeteries into 4 large ones, hence this large cemetery with more than 44,000 dead. It includes pillboxes that were once part of the German frontline.
N50 55.197 E2 55.017 (50.91995 2.91695)

2. **KITCHENER'S WOODS:** On April 22, two Canadian battalions, the 10th of the 2nd Brigade and the 16th of the 3rd Brigade, at 2300 launched a counterattack into the gap created by the German gas attack. The Canadians charged 180 meters through a hedge and into devastating fire from German machine gunners lying in the woods. Because of their heroism, the German attack was stopped, but they suffered a casualty rate of more than 70 percent. Located here is a small stone monument in the shape of the old oak trees that once grew here.
N50 53.243 E2 55.137 (50.887383 2.91895)

3. **SAINT JULIAN MEMORIAL, ALSO CALLED VANCOUVER CORNER, "BROODING SOLDIER":** It is here that the Canadians held their lines against the first German gas attacks on the Western Front from April 22-24, during the 2nd Battle of Ypres. The Canadians suffered 6,037 casualties. Today, there is a memorial with stone towers and a statue of a large-jawed, stoic Canadian soldier "brooding" in defense, facing where the German attack came from.
N 50 53.967 E 2 56.433 (50.89945 2.94055)

4. **TOTEN MUHLE:** Located just east of the Saint Julian Memorial, it was an old mill the Germans used as a forward artillery observation post.
N50 53.851 E2 56.902 (50.89945 2.94055)

5. **ARTILLERY WOOD CEMETERY:** This was a front-line cemetery location in 1917. The Irish Poet, Francis Ledwidge, has a special monument commemorating his poem "Soliloquy," just south of the cemetery on the corner of Ledwigeped and Poezelstraat. He is buried here along with 1,036 other heroes. To note, the German author and soldier, Ernest Junger, also fought in the battle for this ridge, and wrote about it in his book, "Storm of Steel."
N50 53.975 E2 52.330 (50.899583 2.872167)

YPRES – MIDDLE SECTION/MENIN ROAD TO THE EAST:

6. **YPRES FLANDERS FIELD MUSEUM (LAKENHALLE CLOTH HALL):** In the center of Ypres in the beautifully rebuilt and famous cloth hall is the Flanders Field Museum. The downtown is a "must see" for a meal and a beer in the town

square. It is well-done and recently updated.
N50 51.061 E2 53.126 (50.851017 2.885433)

7. **MENIN GATE:** This monument was built to honor the more than 55,000 Commonwealth soldiers who died during the war, but whose remains were never found. The location was picked because it was the gate through which thousands of the soldiers marched to their trench lines to face the Germans. Here, a very special event, The Last Post, a tradition to honor the veterans and dead, was started May 1, 1928, by the Ypres police. Every day at 8 p.m., the gate road is shut and 2 to 6 buglers arrive and play "the Last Post." It has been played every night since 1928, disrupted only during WW2. Parking is available in the square or outside the city wall.
N50 51.120 E2 53.457 (50.852 2.89095)

8. **HELLFIRE CORNER:** Not too far from Menin Gate on the Menin Road toward the German lines was a location considered "Hellfire Corner" because of all the artillery it attracted on the main march to the front lines. It also indicates the Germans' farthest point of advancement. Off the northeast side of the roundabout is a small marker.
N50 50.937 E2 54.989 (50.84895 2.916483)

9. **HOOGE CRATER MUSEUM:** In this location is a well-done private museum and restaurant located at the old chapel of Hooge, and was a frontline location. The museum has a Fokker DR1 (Red Baron) aircraft hanging from the ceiling and a great collection of other artifacts, weapons, and uniforms. Please note that across the street is the British Hooge Cemetery with 5,500 graves, and directly behind the cemetery is Zouave Wood and Sanctuary Wood.
N50 50.790 E2 56.597 (50.8465 2.943283)

10. **HOOGE CHATEAU/CRATER:** Just a 2-minute walk from the museum is an open air "museum/walk" to the crater site where on July 19, 1915, the Allies detonated 3,500 pounds of ammonal under the German lines. This is definitely worth a visit; just a small donation and open the gate to stroll around to see the former trenches, artillery shells, bunkers, and craters.
N50 50.776 E2 56.745 (50.846267 2.94575)

11. **SANCTUARY WOOD CEMETERY:** Just south of Hooge, on the front lines. It was named by the British soldiers who used it to shelter from the battle during 1914. It was a location of intense fighting between the Canadians and the Germans during 1915. After the war, the British consolidated 637 graves in this location.
N50 50.297 E2 56.650 (50.838283 2.944167)

12. **SANCTUARY WOOD MUSEUM, HILL 62:** Privately owned, you can tour trenches that were actually support/communication trenches for the Allies. There is also material available to buy from WW1 and WW2.
N50 50.214 E2 56.757 (50.8369 2.94595)

13. **HILL 62, CANADIAN MONUMENT:** From April to August, 1916, the Canadians had their first independent operation, fighting over Hill 62 and Mount Sorrel, small hills that provided the Germans some height over the Allied lines. The Canadians gained and lost the dominant hill, but finally pushed the Germans out, thus blocking their clear view of Ypres. The Canadians suffered 8,430 casualties during the fighting, the Germans approximately 4,000.
N50 50.065 E2 56.836 (50.834417, 2.947267)

14. **GERMAN LINES/BUNKERS:** This geo coordinate is at a location for parking, from which you can go into the woods and explore the remains of German trenches and bunkers.
N50 49.848 E2 57.122 (50.8308, 2.952033)

PASSCHENDAELE/ZONNEBEKE (3RD BATTLE OF YPRES) EAST OF YPRES

15. **POLYGON WOOD CEMETERY, BUTTE CEMETERY, AND AUSTRALIAN 5TH DIVISION MEMORIAL:** This is where the British held the Germans back in the 1st Ypres battle, later losing it for 2 years. In the 3rd one, the Australians took back Polygon Woods. At this location is an Australian memorial in Butte Cemetery that holds more than 370 graves, and across the street is the Polygon Wood Cemetery, with 87 graves.
N50 51.428 E2 59.436 (N50 51.428, E2 59.436)

16. **TYNE COT CEMETERY & VISITOR CENTER:** This is the largest British military cemetery in the world, with more than 12,000 buried and 35,000 names of the missing. It is located where front-line German pillboxes stood and has good examples of the German system of defense, with five pillboxes providing supporting fire. During the 3rd Battle of Ypres, the Allies overcame the bunkers during the attack on Passchendaele.
N50 53.259 E3 00.063 (50.88765, 3.00105)

17. **CREST FARM CANADIAN MEMORIAL:** This is a great spot to see the approach to Passchendaele and a very important memorial. From this location, you can see the Canadian path of attack, where they crossed the valley to the east toward the town in October-November, 1917. Their objective was to capture the town center and the dominating ridge. In 16 brutal days, the Canadians suffered 16,000 casualties to gain only a half a kilometer of Passchendaele Ridge, from Crest Farm to the town center. Also note that in the church is a stained-glass window as a memorial (back right).
N50 53.866 E3 00.794 (50.897767, 3.013233)

18. **MUSEUM PASSCHENDAELE-ZONNEBEKE:** Best museum in the area. It provides an excellent overview of all the battles and has a great display of materiel and even an exposition of the tunneling that took place.
N 50 52.236 E 2 59.248 (50.870603, 2.987475)

SOUTH OF MENIN ROAD AND BATTLE OF MESSINES RIDGE

19. **HILL 60 MEMORIAL AND CATERPILLAR CRATER:** This is one my favorite spots in the area. You can explore destroyed German lines and the terrain that was fought over. Just across the railway tracks there is a path to the Caterpillar Crater, one of the large craters left, and created by a detonation of mines under the German lines. Also, don't miss a sign across the road on the bridge from the path to the crater. It has a great information board showing the panorama of the battle field.
    - A. **HILL 60 (GATE JUST TO THE LEFT OF PARKING):**
      N50 49.445 E2 55.678 (50.824083, 2.927967)
    - B. **INFORMATION BOARD WITH PANORAMA:**
      N50 49.426 E2 55.639 (50.823767, 2.927317)
      (It's located across the street and over the railroad bridge)
    - C. **CATERPILLAR CRATER:**
      N50 49.351 E2 55.736 (50.822517, 2.928933)

20. **BEDFORD HOUSE CEMETERY:** This location is just a little to the south, just outside of Ypres. It was the location of a manor house that was turned into a front-line aid station and then into a cemetery with 5,000-plus Commonwealth graves. It received direct fire from German weapons, such as 8-inch cannons, and in one day was targeted by 500 gas shells.
    N50 49.623 E2 53.295 (50.82705 2.88825)

21. **ST. ELOOI MEMORIAL, KRUPP CANNON:** The German-made cannon marks the British Line, and just to its left is a path to two craters with information boards about the battle here in March, 1916, and June, 1917.
    N 50 48.614 E 2 53.494 (50.810233, 2.891567)

22. **BAYERNWALD WOODS GERMAN TRENCHES:** Great place to see part of the German trench system. But, it requires a code to get into the fenced area, which you must get beforehand from the Kemmel Tourist office.
    - A. **BAYERNWALD:**
      N 50 48.090 E 2 52.614 (50.801500, 2.876900)
    - B. **KEMMEL TOURIST OFFICE(FOR THE CODE):**
      N 50 47.022 E 2 49.642 (50.783701, 2.827365)

23. **AMERICAN KEMMEL MONUMENT:** Dedicated to the 27th and 30th US Infantry Divisions. In 1918, they launched an attack on the German trench lines from this point as part of the 5th Battle of Ypres.
    N 50 47.866 E 2 50.939 (50.797769, 2.848983)

24. **16TH IRISH DIVISION MEMORIAL (CELTIC CROSS) AND WYTSHAETE COMMONWEALTH CEMETERY:** At this location is a memorial to the Irish 16th ID that fought in the battle for Messines Ridge, and a Commonwealth cemetery with 1,002 graves. The town was captured on June 7, 1917. It was recaptured

by the Germans, however, in their Spring Offensive of April 16, 1918. Finally, it was recaptured by the Allies on September 28, 1918.
N50 47.070 E2 52.601 (50.784513, 2.876896)

25. **POOL OF PEACE, MINE CRATER, AND LONE TREE COMMONWEALTH CEMETERY:** The Pool of Peace is, in fact, a huge pond created by a crater from one of the 19 mines blown in the Battle for Messines Ridge in June, 1917. Across the street is Lone Tree Commonwealth Cemetery and memorial holding the graves of 88 soldiers who fell, most of them on the first day of the battle for Messines Ridge.
N50 46.540 E 2 51.678 (50.775663, 2.861305)

26. **DE SINT NIKLAASKERK:** The Germans had an aid station in the crypt of Saint Nicholas Church in Messines. It is rumored to be where Hitler was treated while a soldier in WW1. Today, you can visit the church and crypt, which has a memorial to the sacrifices of the war.
N 50 45.837 E 2 53.927 (50.763980, 2.898839)

27. **ISLE OF IRELAND PEACE PARK AND NEW ZEALAND MEMORIAL MESSINES:** The Peace Park commemorates the Protestant and Catholic Irish soldiers who fought together and died here. We have to remember that Ireland was in the middle of its rebellion against the English at this time. It is a really impressive memorial with a newly built Irish round tower and stone markers with Irish soldier/writer/poet quotes. In the back is a path leading to the outstanding New Zealand Memorial Park (No. 29 below).
N 50 45.589 E 2 53.739 (50.759814, 2.895647)

28. **BRITISH CEMETERY ON MESSINES RIDGE AND NEW ZEALAND MEMORIAL TO THE MISSING:** The British Cemetery on Messines Ridge contains the graves of 577 soldiers, the majority of whom died in the battle for Messines Ridge. At the entrance to the cemetery is a cross and a raised platform with the names of 828 New Zealand soldiers whose graves remain unknown.
N 50 45.917 E 2 53.440 (50.765286, 2.890673)

29. **29.NEW ZEALAND MEMORIAL PARK ON MESSINES RIDGE:** Great memorial, because it has remains of German bunkers. To get access to the park, you can leave your car at the Irish Park and walk up a path connecting the two.
N 50 45.642 E 2 53.459(50.760693, 2.890978)

30. **MEMORIAL TO THE CHRISTMAS TRUCE, KHAKI CHUMS CROSS:** At this location is a wooden cross and information board to commemorate the Christmas Day Truce, when the Germans and Allies played a game of soccer on Christmas Day, 1914.
N 50 44.666 E 2 54.148(50.744426, 2.902471)

FARTHER EAST:

31. **AMERICAN FLANDERS FIELD CEMETERY:** This is the only WW1 American Cemetery in Belgium, and has more than 400 graves. It is an extremely beautiful

and an honorable resting place for the American warriors.
N 50 52.415 E 3 27.135(50.873585, 3.452257)

32. **AMERICAN MONUMENT – AUDENARDE (ALSO SPELLED OUDENAARDE):** From October 30 to November 11, 1918, the American 37th and 91st IDs fought under the command of Belgian King Albert I. They were part of the Allied attack that crossed the Scheldt River. The memorial built by the American Monuments Commission is located downtown, across from the main parking lot. The town is a nice stop for lunch and a cold Belgium beer.
N 50 50.790E 3 36.155 (50.8465 3.602583)

NORTH NEARBY IN BELGIUM:

33. **DE IJZERTOREN, DIKSMUIDE, BELGIUM:** It is a tower built to Flemish liberty and to commemorate the Belgian fighting that took place here in WW1. It is also a pilgrimage site for the Flemish people, who make up 60 percent of the country's population (mostly in the north). The remainder of the country is considered Walloon (French speaking), typically located in the south. During WW1, most of the soldiers were Flemish-speaking, but had to take commands in French; this led to formal calls for recognition, and tensions over it continue to the present day. To note, economically the Flemish were historically the poor farmers but today they have gained in prosperity as land owners and in the farming trade, which creates different dynamics than in the past.
N 51 01.900 E 2 51.126 (51.031666, 2.852102)

34. **BATTLE OF YSER, TRENCH OF DEATH:** Just north of Diksmuide on the Yser Canal lies this trench line, preserved with concrete, and which became the Belgian main line of defense against the Germans. Early in the war, the 2 sides "raced to the sea," trying to outflank one another. After sharp and intense battles, the front, although still miserable and deadly, stabilized to an extent for the rest of the war.
N51 2.511 E2 50.871 (51.04185 2.84785)

35. **EDITH CAVELL:** "Standing as I do in view of God and eternity, I realize that patriotism is not enough. I must have no hatred or bitterness towards anyone." Edith wrote that passage in her diary the day before her execution on October 12, 1915.

Edith Cavell was born December 4, 1865, in Norfolk, England, and throughout her life she was dedicated to helping others. She became a nurse in England and later in Belgium. She bravely broke through barriers regarding women earning a living, including in the nursing profession. In Belgium, she became a leader in instructing women in nursing prior to the war.

The war broke out while she was in England, but against the wishes of her family and friends, she knew her duty was in Belgium. Once there, she took care of wounded soldiers from all nations. When Brussels was captured, most of the English nurses were sent home and replaced by German ones but she stayed. During the battle for Belgium, many British and Allied soldiers were trapped behind enemy lines and over the course of a year, Edith secretly assisted approximately 200 of them in escaping to the neutral Netherlands or back to the Allied lines. She was eventually denounced by a civilian collaborator and arrested on August 3, 1915 by the Germans. She confessed to the Germans and stood by her actions as the just and right thing to do, even though it was against the rules of war.

Edith was held in prison for 10 weeks. Tragically, despite diplomatic requests from the highest levels of many governments, she was executed by a German firing squad on October 12, 1915. There was outrage at her death and it was "used" as propaganda against the Germans.

Edith is remembered today for her heroism, humility, and love, and for her work in modernizing nursing in Belgium. A memorial in her memory is in Brussels at the geo coordinates below outside a hospital. It depicts two cast angels pointing at her name, the text reading, "Passerby, tell your children that they killed them." It says "killed them" because the memorial is also for Marie Depage, a nurse killed on the Lusitania after a trip to the USA to raise funds for hospitals.
N 50 48.791 E 4 21.411 (50.813187, 4.356849)

36. **GERMAN CEMETERY VLADSLO, BELGIUM:** At this location are buried more than 25,000 German soldiers from WW1. The cemetery is also famous for its statue of "Grieving Parents," created by the German artist Kathe Kollwitz, whose son is buried in the cemetery. It is located just about 8 km east of Diksmuide.
N 51 04.242 E 2 55.762 (51.070703, 2.929372)

37. **KOEKELARE, LONG MAX SITE, AND MUSEUM:** During WW1, the Germans used large siege cannons behind the front lines, built into concrete emplacements. Later in the war, they used rail guns because of the time required to build the concrete emplacements. Here, a new museum has been built dedicated to a 380mm German siege gun "Long Max." The museum is very informative, new, and you can explore the remains of the Long Max emplacement. It is not far from Vladslo.
N 51 07.013 E 2 58.942 (51.116891, 2.982371)

THE FIVE BATTLES OF YPRES, NOVEMBER 1914 TO OCTOBER 1918

MAP 7

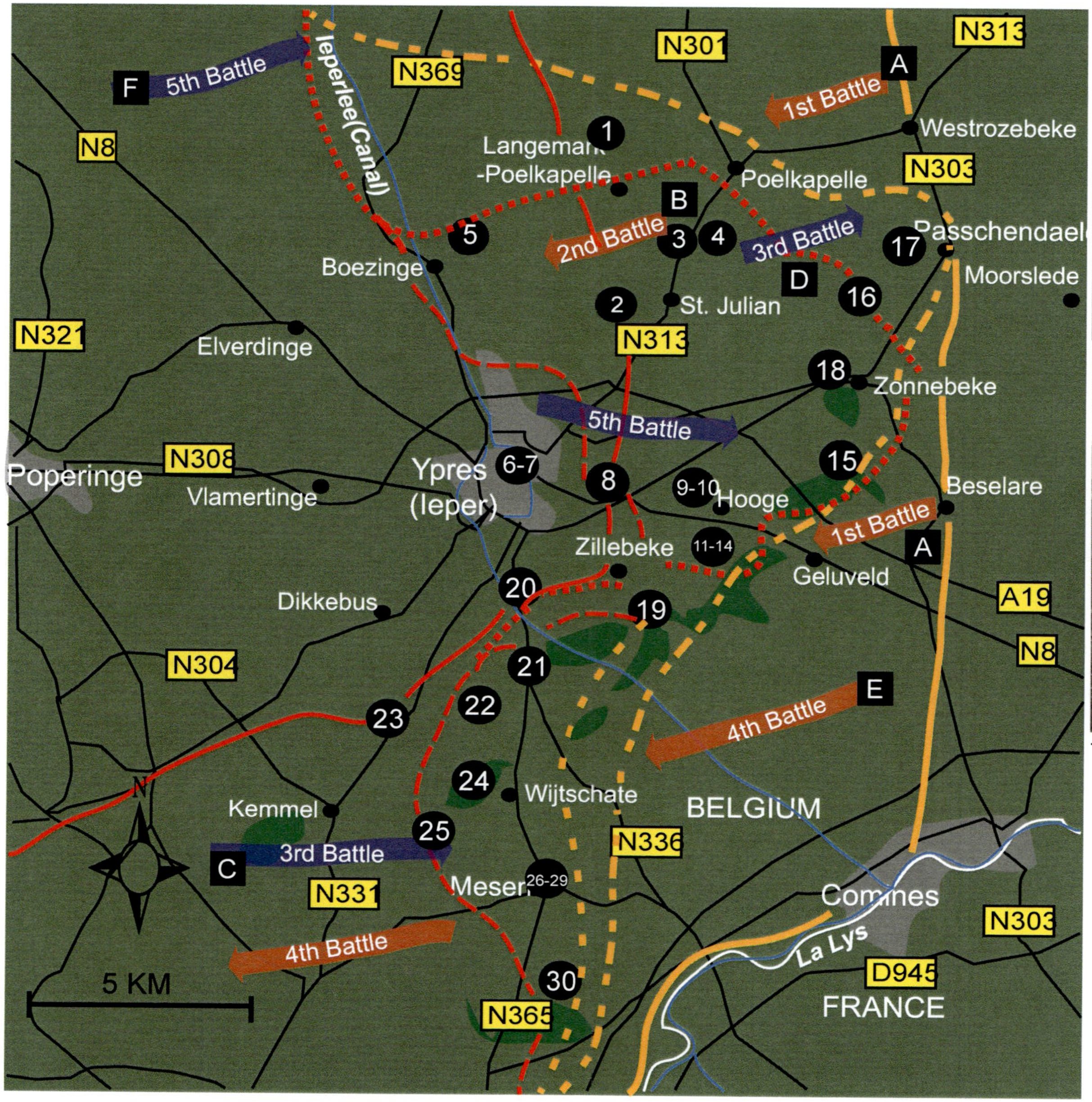

KEY:
Main thrust, please note the entire battle typically took up the entire front
Nov 10, 1914, 1st Ypres Line
May 25, 1915, 2nd Ypres Line
June 14, 1917 3rd Ypres, Messine Ridge
Nov 10, 1917, 3rd Ypres, Passchendaele
April 30, 1918, 4th Ypres
Sept 28 - Oct 2, 1918, 5th Ypres

The famous Cloth Hall in Ypres., September 1918.

THE FIVE BATTLES OF YPRES, NOVEMBER 1914 TO OCTOBER 1918

MAP 7

Movements:

1st Battle of Ypres, October 19 - November 22, 1914
A. The Belgians, British and French stopped the Germans from capturing the cross roads town of Ypres. But the Germans captured the high ground of Messines Ridge.
2nd Battle of Ypres, April 22- May 15, 1915
B. The Germans used chemical weapons for the first time. The Canadians filled the line and held the Germans from a total break through.

3rd Battle of Ypres, Messine Ridge June 7-14, 1917
C. South of Ypres the British had a very successful operation to capture Messine Ridge, with the explosion of 19 major mines.
3rd Battle of Ypres, Passchendaele, July 31 - November 10, 1917
D. After delaying their Operation in the north, the British grinded their soldiers into the German line, in one of the costliest battles of the war, known for the fight for Paschendaele.

4th Battle of Ypres, German Spring Offensive, April 9-25, 1918
E. The Germans attacked their main thrust south of Ypres with some gains, the goal to split the British and French. Within 2 weeks their momentum was lost.
5th Battle of Ypres, Grand Offensive/Lys Offensive, Sept 28 - Oct 2, 1918
F. The Belgian King, led the major counterattack in the north and from Ypres, with decent gains in terroritory, until supplies dwindled and German enforcements arrived.

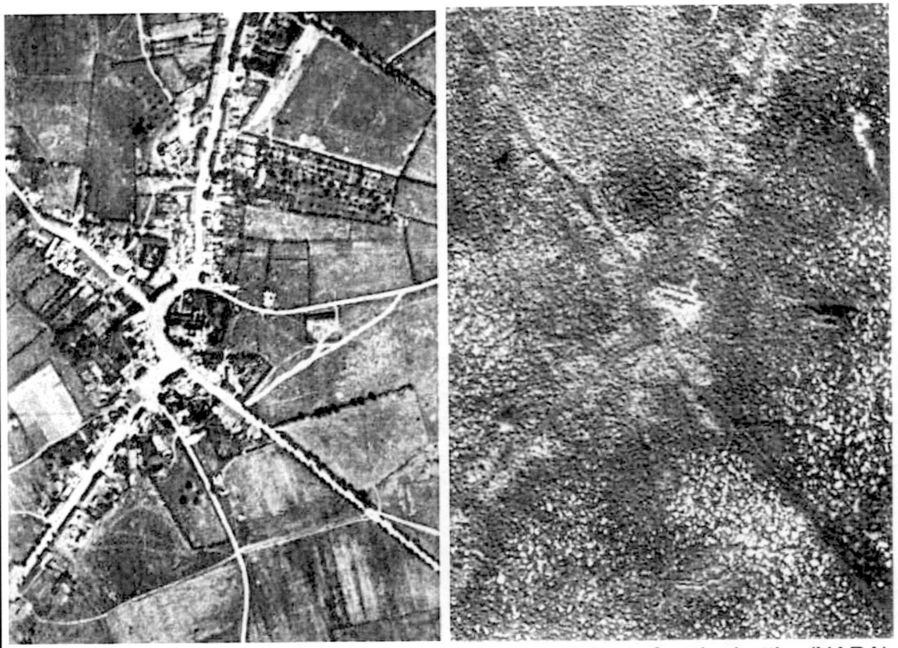

Photo above is Passchendaele before the battle and then after the battle. (NARA)

Places to See:

Northern Section:
1 Langemark German Cemetry
2 Kitchner's Woods
3 St. Julien Canadian Memoria
4 Toten Muhle
5 Artillery Wood Cemetery

Middle Section/Ypres:
6 Ypres Lakenhalle Museum
7 Menin Gate
8 Hellfire Corner
9 Hooge Museum
10 Hooge Crater
11 Sanctuary Wood Cemetery
12 Sanctuary Wood Museum
13 Hill 62, Canadian Monument
14 German Lines/Bunker Remains

Passchendaele Locations
15 Polygon Wood Cmtry & Memrl
16 Tyne Cot Cemetery
17 Crest Farm Memorial
18 Museum Passchendaele

Southern Section/Messines Ridge:
19 Hill 60 & Caterpillar Crater
20 Bedford House Cemetery
21 St. Elooi, Krupp Cannon
22 Bayernwood Trenches
23 American Kemmel Memrl
24 Irish 16th ID Memorial
25 Pool of Peace Memorial
26 De Sint Niklasskerk
27 Isle of Ireland Peace Park
28 British Cemetery Messine Ridge
29 New Zealand Memorial Park
30 Christmas Truce Tree Memorial

Ypres or Iepers, has been rebuilt from the devastation in WW1. Above left is the famous Cloth Hall which now houses a museum. Above right is a photo of Menin Gate, where the names of over 55,000 Commonwealth soldiers are engraved who died and were never found. Menin Gate was the gate which many of the soldiers marched through towards the east and the German lines. Every night, they play "The Last Post" and on some occasions like above have dignitaries recite a poem and lay a wreath.

Tyne Cot cemetery is the largest Commonwealth Cemetery in the world with over 12,000 buried soldiers and over 35,000 names of the missing. It is located on the German front lines, which were captured by the Allies. Uniquely five German bunkers still remain in a pattern which allows the visitor to understand how deadly the supporting cross fire could be.

Above left is a view of St. Nicholas Church in Messines on Messines Ridge. The dominating ridge south of Ypres was the scene of terrible fighting through the war. Today you can tour the quiet roads linking the many memorials, as seen in photo above right.

Above statue is a memorial to the many diggers or miners that the Allies relied on to dig mines under the German lin esespecially under Messines Ridge.

Above left is one of the many German bunkers throughout the battlefields and an explanation board. This one is near Messines ridge. The photo right is of the Irish Peace Park, a memorial to the Catholic and Protestant Irish who fought for the Commonwealth during WW1. As part of the memorial, they built a famous Irish Round Tower and quotes from soldiers.

The lines around Hill 60, were the scene of vicious fighting. Today many craters and a torn landscape remains to explore. Nearby next to the Hill 60 memorial is Caterpillar Crater, one of the largest mine detonations, shown in photo above right.

Above left is the Canadian Crest Farm Memorial facing the church steeple of Passchendaele, only half a kilometer away. In 1917 the Canadians fought for 16 days to capture the town and suffered over 16,000 casualties. Photo right is of the Zonnebeke Museum on the Ypres battles, it is my favorite for this battlefield and located only minutes from Crest Farm.

The Brooding Soldier, is the Saint Julien Wood Memorial, this is the location the Canadians held the line against the German gas attacks in April 1915. Photo right, throughout Flanders Fields you will find the poppies "row on row". Legend has it that they used to be white until the arrival of Attila the Hun who decimated Europe in the 400s and he spilled so much blood the poppies became red

Photo left, is of the German Cemetery, Langemark, which has bunkers preserved within the cemetery. It holds the remains of over 44,000 German soldiers.

Photo left is from the Canadian Memorial, Hill 62. Although only 62 meters above sea level, the slight gain in elevation provided the Germans better visibility of Ypres, which you can see on the horizon. The Canadians suffered over 8,000 casualties taking the hill. Photo right is of Hooge Crater, which like most of the craters from mine explosions are now ponds. Hooge has a great walking path with artifacts and trench remains all around. Just look out for the very large sheep!

The above photo is of the Australian Memorial to the 5th Australian Division who held the "Butte" which was a man made land mass used by the Belgians for a firing range. Today the cemetery and memorial are surrounded by new growth trees but after the battle see photo upper right of the remains of the forest and Butte.(NARA) Photo right is of the American memorial in Kemmel looking east towards the German lines. The Americans joined the battle in 1918 when they were part of a major counterattack.

BATTLE OF ARRAS

October 1-4, 1914

(Battles of Artois)
First, December 17, 1914-January
Second, May 9-June 19, 1914
Third, September 25-28, 1915

The battlefield "resembled a furious sea...the image of the earth after the flood"—Galthier-Boissiere

---

SUMMARY: The Battle of Arras was located between the battlefields of Ypres to the north and the Somme River directly south. As with the entire front, the Germans attempted to break through at this location, which also became known as the Battles of Artois, the region that Arras is part of. The terrain is generally flat except for ridgelines that dominate the surrounding area. The ridges were critical because they guarded the Douai Plain, which was a rich source of French coal and manufacturing.

In the initial weeks of the war, September, 1914, the Germans were on the eastern outskirts of Arras and about to capture the city until the French 77th Mountain Division commanded by Gen. Ernest Jaques Barbot arrived and stubbornly held the city through heavy German bombardments well into October. After several attempts to attack the city during the Battle of Arras, the Germans went around it to the north and captured the important towns and ridges, such as Vimy, Notre Dame de Lorette, St. Nazaire, and Givenchy. After they held the high ground, they focused on trying to outflank the Allies until they hit the English Channel. That led to the Battle of Yser and Battles of Ypres. The following is a brief summary of the battles that took place. The battles overlapped in time and geography; as a result so do the area memorials.

The First Battle of Artois is considered to have begun at 1300 on December 17, 1914, one of the first major Allied offensives of the war. The French pounded the German positions on Notre Dame Lorette for 6 hours, then attacked to capture the high ground. The battle ended in failure and extreme casualties for the French.

The Second Battle of Artois began on May 9, 1915 with the goal of capturing the high ground around Arras. After more than 40 days of battle and 140,000 casualties, the French recaptured the high ground of St. Nazaire, Notre Dame Lorette, and Neuville St. Vaast.

In the Third Battle of Artois, the French attacked on September 25, 1915, while the British began their attack for the Battle of Loos just north of Arras. The French successfully straightened out their lines and captured additional high ground of Souchez and pushed the Germans back to Vimy Ridge. The French could go no farther, however. After three weeks and gaining no more than 2 km, they had suffered more than 135,000 casualties, the Germans more than 170,000.

OUTCOME: After the battles of 1915, the area stabilized and no new offenses were initiated here until April, 1917 when the Canadians captured Vimy Ridge.

PLACES TO SEE: It starts with the number 12 because this shares a map with the Battle of Vimy Ridge, please see that map for the locations of the "Places to See."

12. **MEMORIAL DE NOTRE DAME DE LORETTE, RING OF REMEMBRANCE AND MUSEUM VIVANT:** The Memorial de Notre Dame de Lorette holds the remains of more than 44,000 French soldiers. It was the high ground fought over during the Battles of Artois and Arras. The memorial contains a beautiful basilica and has an incredible view of Vimy Ridge and the Douai Plain. The French lost it early in 1914 and recaptured it at great cost in 1915. General Barbot, the defender of Arras, is buried here. The location also includes the Lanterne des Morts, which can be seen at night more than 60 km away. The location also has Musee Vivant, 1914-18. Also for the centennial, they built a memorial to 540,000 soldiers who died in the Nord and Pas-de-Calais departments. The memorial lists in alphabetical all of their names, of all ranks and all nations. It is an incredible and bone-chilling memorial. It is located just across the street from Notre Dame de Lorette.
N 50 24.057 E 2 43.145 (50.40095, 2.719083)

13. **CABARET ROUGE CEMETERY:** At this location is a large and well-kept Commonwealth Cemetery. It contains more than 7,650 burials of WWI. The name comes from a café that was there, but was destroyed during 1915 artillery bombardment.
N 50 22.839 E 2 44.495 (50.380650, 2.741576)

14. **MEMORIAL TO POLISH VOLUNTEERS AND MEMORIAL TO CZECHOSLOVAKIA VOLUNTEERS:** At this location is a memorial to the Polish who fought for the French during WWI, and across the road is a memorial to the Czechoslovakia volunteers.
N 50 21.937 E 2 44.701 (50.365624, 2.745012)

15. **LICHFIELD CRATER:** The Canadians used this crater as a mass grave for the remains of approximately 60 Canadian soldiers killed April 9-10, 1917.
N 50 21.571 E 2 46.638 (50.359517, 2.7773)

16. **GERMAN CEMETERY, ARRAS:** This German cemetery is the largest in France, with more than 44,000 graves.
N 50 20.530 E 2 45.251 (50.342167, 2.754183)

17. **GENERAL BARBOT AND 77TH DIVISION MONUMENT:** In the town of Souchez is a memorial to General Barbot of the 77th Mountain Division. In 1914 against orders, he courageously held against overwhelming German attacks. Because of his courage and leadership, he kept the Germans from capturing Arras. Barbot died of wounds in May, 1915. His replacement, Colonel Jean Paul Ernest Stirn, also died two days later, and his name is also memorialized on the monument.
N 50 23.177 E 2 44.661 (50.386280, 2.744343)

18. **LENS CENTER OF PEACE, 1914-1918:** Opened in 2015, the free exhibit documents the battles that took place in the region. It has excellent modern displays, maps and documentation. Very well done.
N 50 24.068 E 2 44.408 (50.401127, 2.740125)

Above photo is of the Memorial and Chapel, Notre Dame de Lorette, St. Nazaire. Located on the ridge west of Vimy Ridge. In 1914-15 the French fought gallanty recapture and hold it. Below photo is of the original chapel on the ridge in 1915, photo right is a view from the memorial with Vimy Ridge in the distance.

Photo left is the inside of "The Ring of Rememberance", containing 540,000 names of all ranks and nations that fell in this region of the war, a gut wrenching walk. Photo right, the French General Barbot memorial, defender of Arras.

FIRST BATTLE OF CHAMPAGNE

December 20, 1914-March 17, 1915

SECOND BATTLE OF CHAMPAGNE

September 22-October 6, 1915

"Our experience has proved that one can carry the whole defensive system in a single bound."—Joseph Joffre

---

SUMMARY: After the "Miracle on the Marne," both sides settled into defensive positions. French Commander Joseph Joffre wanted a decisive battle to end the war within weeks. His plan was to stage a major offensive attacking the northern and southern edges of the Sayon Salient, which was held by the German 3rd Army. Joffre believed an overwhelming force would break the German lines and envelop them. Initially, the French attacked in the south, in Champagne at Perthes and Noyon. Then the British attacked in the north at Givenchy, near Nieuwpoort, Belgium. The Germans' defensive line repulsed most of the attacks and fighting continued until March, when the French attack was called off. The French gained only 3 km at the cost of 90,000 casualties. The Germans suffered about the same number.

On September 22, 1915, Joffre, convinced that the Germans were still weak and spread thin, decided to attack once again in what became the Second Battle of Champagne. The British attacked in the north again, instigating the Battle of Loos. This time, Joffre gathered artillery en masse and for 3 days bombarded the German lines in the Champagne region. Then the infantry attacked. The effort was successful at first, gaining 3 km, but German reinforcements arrived and pushed the French back. The result was tremendous losses for the French, with little area gained and a year wasted. The Germans were able to strengthen their positions and weaken the French forces.

OUTCOME: The Germans held their ground and defeated the Allied offensive. In the Second Battle of Champagne, the French pulled artillery from the Verdun region for their bombardment and offensive. The Germans took note of Verdun's weakened state and knew the French had suffered many casualties. Consequently, the Germans planned a February, 1916, attack on Verdun with the goal of bleeding France white.

WHO? The British, French, and Germans. The Allies lost 90,000 men in the first battle and more than 145,000 in the second. The Germans suffered 90,000 casualties in the first battle and approximately 72,500 in the second. It was an astonishing amount of casualties for battles that lasted a total of 5 months.

WHEN? The First Battle lasted from December 20, 1914, through March 17, 1915. The Second Battle lasted September 22 through October 6, 1915.

WHY AT THESE LOCATIONS? The French believed the Champagne Salient was the weakest point in the German lines.

TECHNOLOGY/TACTICS: Technology was ahead of tactics. The Germans built up and supplemented their defenses through the use of machine guns, barbed wire, and mines. The French were still developing tactics, as evidenced by the fact they used a massive artillery bombardment during the Second Battle in an effort to "open up" the German lines.

PLACES TO SEE:

1. **FORT DE LA POMPELLE:** This is a fort outside of Reims, from which the French fought. It held its own throughout the war and has the battle scars to prove it. It is a very excellent exhibit and tour, with easy access from the main highway.
   N 49 13.072 E 4 07.750 (49.217867, 4.129167)

2. **MONUMENT OSSUAIRE DE LA FERME DE NAVARIN:** Off the main highway sits a memorial that commemorates 10,000 warriors who died fighting for France's freedom, most of them during the Battle of Champagne. Its location was the epicenter of the front. Three statues of soldiers advancing to battle top the pyramid-shaped tomb. One of the statues has the features of Quentin Roosevelt a pilot and the son of Theodore Roosevelt, who died on July 14, 1917, in Cambrai. The monument is located at the intersection of D220 and D977, and only 300 meters north is a German pillbox.
   N 49 13.115 E 4 32.542 (49.218585, 4.542364)

3. **MUSEUM, MARNE 14-18, SUIPPES:** Located just south of the Navarin Memorial is a newly renovated museum with excellent displays, interactive trenches, and new interpretations.
   N 49 07.944 E 4 31.924 (49.1324, 4.532067)

4. **FRENCH CEMETERY, SUIPPES:** Located here is a very large French cemetery with approximately 4,000 graves. It sits on the northeast corner of the town of Suippes and shares land with a mass grave containing 5,539 German soldiers.
   N 49 07.964 E 4 32.105 (49.132733, 4.535083)

5. **FRENCH CEMETERY, JONCHERY-SUR-SUIPPE:** Also located next to the Suippe River is another major French cemetery, which contains the remains of more than 7,000 soldiers.
   N 49 09.153 E 4 28.720 (49.152545, 4.478661)

6. **FRENCH AND GERMAN CEMETERIES, SOUAIN PERTHES LES HURLUS:** Incredibly, a French and a German cemetery lie next to each other. The French cemetery, La Crouee, contains 30,732 graves, the German one 13,000 graves.
N 49 11.243 E 4 32.261 (49.18738, 4.537675)

7. **RUSSIAN CEMETERY AND MEMORIALS, SAINT-HILAIRE-LE-GRAND:** This cemetery contains the graves of more than 900 Russian soldiers who fell while fighting for France during WW1.
N 49 09.492 E 4 23.999 (49.158204, 4.399975)

8. **FRENCH, POLISH AND GERMAN CEMETERY, AUBERIVE:** Here sits another French cemetery next to a German one. The French one contains the graves of 6,424 men who perished during WW1. The German cemetery contains the graves of 5,359 soldiers.
N 49 11.587 E 4 22.084 (49.193119, 4.368062)

**Also nearby is the American memorial, Sommepy, just north of the town of Sommepy. I cover it in the "Grand Offensive" section.*

2ND BATTLE OF CHAMPAGNE, SEPTEMBER 22–OCTOBER 6, 1915

MAP 8

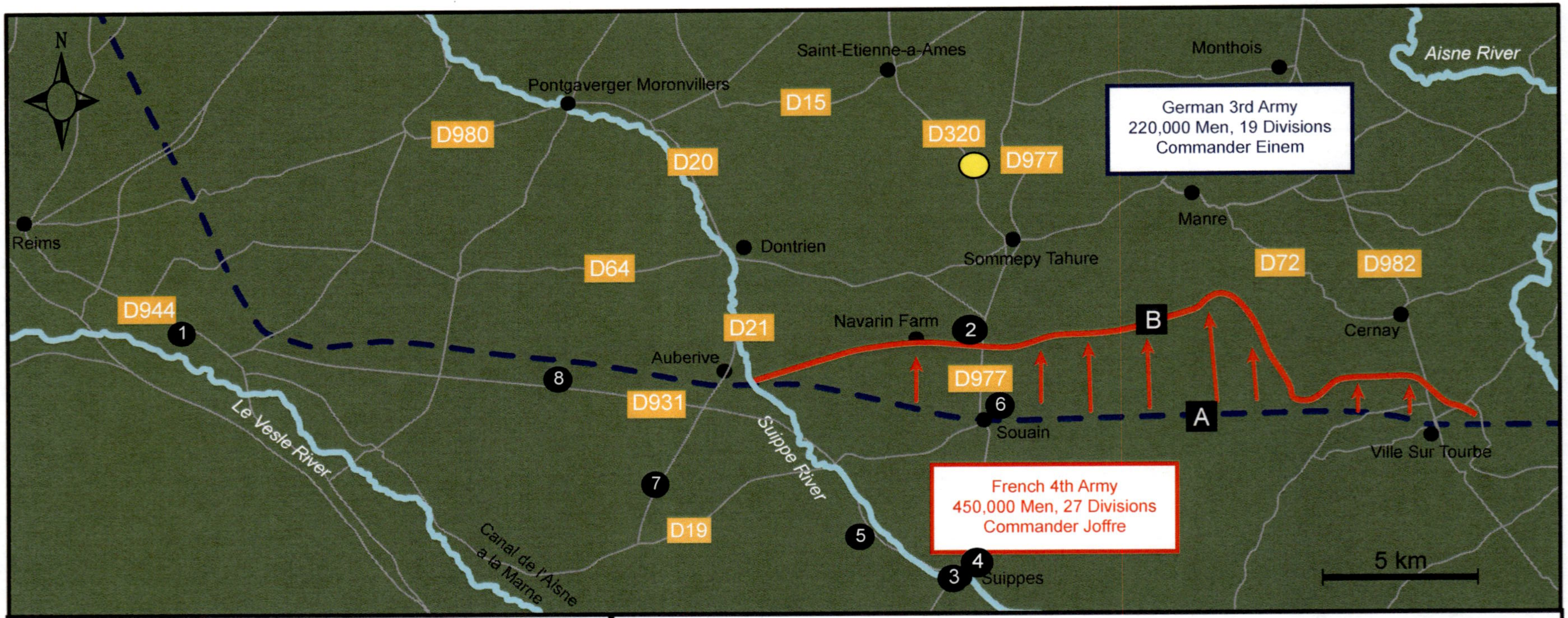

Movements:

A. September 25, after 3 days of artillery bombardments the French Army attacked.

B. By October 6, German reserves stopped any further progress and the battle was over.

Places to See:

1. Fort de la Pompelle
2. Monument ossuaire de la Ferme de Navarin and German Pillbox.
3. Suippes, 14-18 Museum
4. Suippes French Cemetery
5. Jonchery-sur Suippe French Cemetery
6. Souain French and Germany Cemeteries
7. Russian Cemetery, Saint Hilaire le-Grand
8. French and German Cemetery, Auberive

(Yellow circle) American Memorial north of Sommepy, see the Grand Offensive section of this book.

Fort de la Pompelle, outside of Reims, was the cornerstone preventing the Germans from taking the city. The fort took a severe beating during the entire war—and it shows—but the defenders courageously held their ground. Also photo right, the Germans shelled Reims and used the cathedral to range their guns. It was badly damaged during the war but has since been rebuilt.

The cathedral contains many beautiful stained-glass windows, including one (above left) of men crushing grapes for the champagne this area is famous for. The center photo is the Angel of Victory in a city square and the photo to the right is the Smiling Angel, a famous Gothic sculpture (created in the 1220s) know for an optimistic outlook. The angel is smiling for the Annunciation.

La Ferme de Navarino, near Sommepy Tahure, is a memorial and ossuary for 10,000 soldiers who perished in the fields of Champagne. The soldier on the right was carved with the features of Quentin Roosevelt who was shot down on July 14,1918. Photo right is Reims Cathedral during the war getting shelled by the Germans. (NARA)

BATTLE OF LOOS

September 25-October 14, 1915

"Known unto God"—Rudyard Kipling

Summary: The Battle of Loos was one of Britain's first major attacks. The British had recently brought in their new recruits/conscripts to form what was known as Kitchener's Army, named after Lord Herbert Kitchener, the defense minister. At this time, the Germans had settled into a defensive position, and recent assaults by the Allies had failed. In the Battle of Loos, the British sought to attack north of Lens and the French south to oust the Germans from the French mining center. French Commander Joffre also staged a large offensive in Champagne to break through two portions of the German lines. He believed that two major attacks would finally crack the German defense.

At Loos, the British objective was to assault the German lines and penetrate over 2 lines of defense, nearly 7 km. At 0550 on September 25, and after a four-day artillery bombardment, the British opened more than 5,000 cylinders of chlorine gas. Unfortunately, the wind was erratic, and the British forces panicked, stalling the attack and in some cases forcing the British to travel through their own gas. Despite such setbacks, in some areas the British gallantly fought their way through German lines (such as when they captured the significant trench defensive complex known as Hohenzollern Redoubt). During the battle, British Sgt. Frank Edwards kicked a football forward to encourage the men of the Irish Rifles. In many areas, German interlocking machine-gun fire annihilated entire units. The next day, 2 new divisions tried attacking over the flat terrain; they, too, were met by intact barbed wire and withering German machine-gun fire at 1,000 meters. After a lull in fighting, the British lost their momentum and the Germans recaptured Hohenzollern Redoubt. After one more unsuccessful attempt in October, and with bad weather setting in, the British called off the attack.

OUTCOME: Although the British New Army received praise for its bravery, the staff was hammered for its poor planning and incompetence. The commander of the British Expeditionary Forces, Sir John French, was removed shortly afterward. The Germans maintained their lines and strengthened their defenses.

WHO? The British attacked with 6 divisions against an unknown number of Germans. The British suffered more than 50,000 casualties (15,800 killed in action), while the Germans suffered more than 25,000 (1,837 killed in action) in just 3 weeks of fighting.

WHEN? September 25-October 14, 1915

WHY AT THIS LOCATION? The British believed this was a weak point in the German defense and wanted to kick the Germans out of the French mining centers.

TECHNOLOGY/TACTICS: The British used few new tactics. They relied on the predictable pre-bombardment to destroy German gun emplacements and break up barbed wire. They also used chlorine gas for the first time, albeit unsuccessfully. At this time, gas was used not with artillery shells but with canisters that relied on the wind to blow the gas over the battlefield. One can imagine how unreliable such a system was. In their trench system, the Germans used very effective interlocking fields of fire to cover flank attacks and direct advances. They also built effective concrete trenches and strongpoints for protection, defense, and morale.

Rudyard Kipling lost his son in this battle; 18-year-old Jack Kipling was considered missing in action until his grave was found in 1992. His gravestone reads, "A Lieutenant of the Great War Irish Guards." Tests were performed to confirm the identification. Also, Rudyard Kipling coined the epitaph of British unknown soldiers: "Known unto God."

Have You News of my Boy Jack?
By Rudyard Kipling

"Have you news of my boy Jack?"
Not this tide.
"When d'you think that he'll come back?"
Not with this wind blowing, and this tide.

"Has anyone else had word of him?"
Not this tide.
For what is sunk will hardly swim,
Not with this wind blowing, and this tide.

"Oh, dear, what comfort can I find?"
None this tide,
Nor any tide,
Except he did not shame his kind -
Not even with that wind blowing, and that tide.

Then hold your head up all the more,
This tide,
And every tide;
Because he was the son you bore,
And gave to that wind blowing and that tide.

PLACES TO SEE:

1. **DUD CORNER CEMETERY:** This cemetery and memorial are located on what was a German strongpoint, the Lens Road Redoubt. The Scottish 15th Division captured it during the Battle of Loos. It contains the graves of 1,784 British soldiers and 28 Canadian soldiers. The memorial at the back contains the names of 20,584 men who died while fighting in the Battle of Loos (1915) and the Battle of Bethune (1918).
   N 50 27.633 E 2 46.284 (50.46055, 2.7714)

2. **ST. MARY'S ADVANCED DRESSING STATION CEMETERY:** A beautiful path leads to this cemetery located in the middle of a field. It holds the graves of 218 soldiers, including Jack Kipling, the son of Rudyard Kipling. His grave was located in 1992 and bore the descriptor "Unknown Irish Lieutenant." The younger Kipling died during the Battle of Loos and his gravestone is located at Plot VII. D. 2. Also at this location are two additional small but beautiful British cemeteries. The smaller is "Ninth Avenue Cemetery," the other just beyond it Bois Carre Cemetery or "square woods."
N 50 29.137 E 2 47.35 (50.485617, 2.789167)

3. **HOHENZOLLERN REDOUBT AND 46TH ID MEMORIAL:** This was the German strongpoint in the fields. At the beginning of a small dirt road is a memorial commemorating the 46th ID, which assaulted the position during the battle. It is a memorial to the division's bloody battle on October 13, 1915. Attempting to take the redoubt, it suffered more than 3,700 casualties. The coordinate takes you to the German line's strongpoint.
N 50 29.849 E 2 46.620 (50.497483, 2.777)

4. **QUARRY CEMETERY:** This location near Hohenzollern Redoubt is in an old chalk quarry, located in no-man's land during the battle. The first day of the battle, the 9th ID captured the Redoubt, but then it was recaptured by the Germans on October 3.
N 50 29.979 E 2 46.162 (50.499648, 2.769369)

5. **LOOS EN GOHELLE BRITISH CEMETERY:** Here are buried 890 British Commonwealth soldiers. The cemetery shares a wall with the Canadian Hill 70 memorial. The coordinate is for the front entrance of the cemetery.
N 50 27.092 E 2 47.846 (50.451537, 2.797426)

6. **6. HILL 70 CANADIAN MEMORIAL:** After the success of Vimy Ridge in April, 1917, the Canadians were asked again to attack the German line to divert German reserves. This time, the Allied main attack was taking place outside Ypres and getting bogged down. After careful planning, Lt. General Arthur Currie, commander of the Canadian Corps, decided on a surprise attack on the high ground near Loos at Hill 70. The Canadians misled the Germans by conducting artillery fire and trench raids south of Lens, away from Hill 70. On August 15, the Canadians attacked and quickly captured the hill. The Germans counterattacked 21 times! Currie had prepared for the counterattacks by providing his troops extra machine guns to defend the ground with no fewer than 250 machine guns. Next, the Canadians attacked the town of Lens on August 21 and 23. The Canadians suffered significant casualties from the German defensive positions there. The Canadians were only able to capture part of the town before running out of momentum, and the battle was called off on August 25. The short battle was ferocious and caused more than 9,200 casualties out of the force of 100,000 Canadians. The Germans suffered more 25,000 casualties during the 10-day battle. Today, there is a memorial park next to the Loos-en Gohelle cemetery with a memorial to the battle. The coordinate is for parking on the opposite side of the park from the cemetery entrance.
N 50 27.214 E 2 48.144 (50.453569, 2.802405)

BATTLE OF LOOS, SEPTEMBER 25 TO OCTOBER 14, 1915

MAP 9

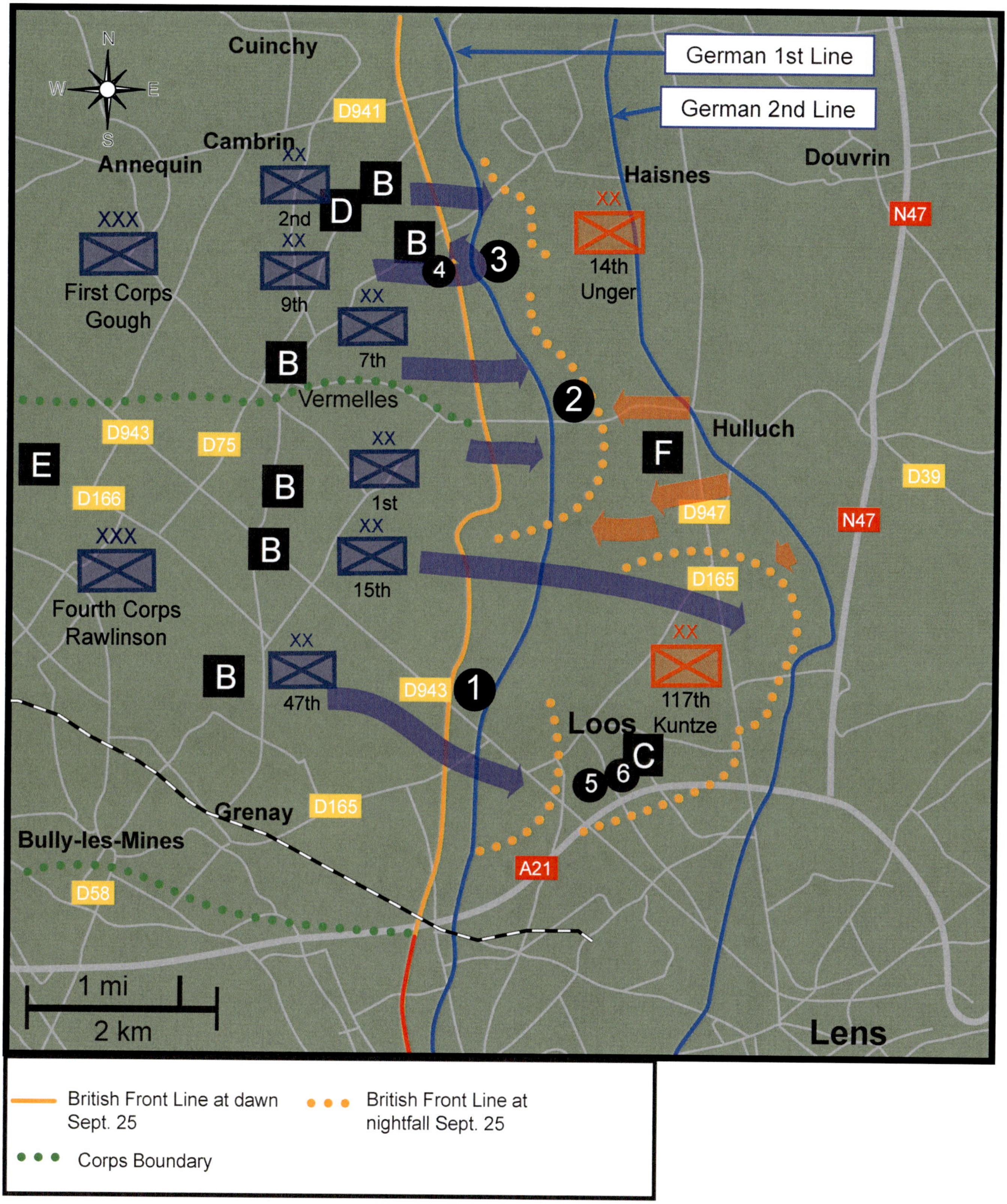

BATTLE OF LOOS, SEPTEMBER 25 TO OCTOBER 14, 1915

MAP 9

MOVEMENTS:

A. At 0550, on Septmber 25, after four days of shelling with over 200 cannons, the British released chlorine gas across the 10 mile front.

B. At 0630 six British Divisions attacked the German line.

C. After an all night street battle, the 15th Division captured Loos.

D.The initial British gas cloud early in the battle effected the British I Corps, resulting in the loss of troops and momentum.

E. Raw British recruits of the 21st and 24th Infantry Divisions arrived a day late, exhausted and unable to assist in breaking through.

F. German counterattacks stopped further breakthroughs and the battle was considered over by October 8, 1915.

PLACES TO SEE:

1. Dud Corner Cemetery
2. St Mary's Advanced Dressing Station
3. Hohenzollern German Redoubt and 46th ID Memorial
4. Quarry Cemetery and Memorial
5. Loos en Gohelle Cemetery
6. Hill 70 Canadian Memorial (From battle in August 1917)

Above, British infantry advance during the Battle of Loos, thru a chemical attack, gun fire and flares. To the left, the remains of a church in Loos after the battle. *(Imperial War Museum)*

These two photos were taken during the Battle of Loos. They show the chemical attacks occuring and the British soldiers slowly advancing through the clouds with their gas masks. (IWM)

Photo left, Dud Corner Cemetery, this photo was taken from the front of the cemetery towards the mine slag mounds. At this locationwas a German strong point taken by the 15th Scottish. Photo right, is St. Mary's Advanced Dressing Station Cemetery. It is located in the middle of a serene and peaceful field today. A beautifully manicured path leads you to the cemetery, where 218 soldiers are buried.

The above photo was taken from another small cemetery just south of St. Mary's ADS looking back at the small "Ninth Avenue, center photo is of John "Jack" Kipling the son of Rudyard Kipling. Jack was killed in action during the Battle of Loos and his grave, right photo was finally confirmed in 1992. The grave is in St. Mary's ADS Cemetery.

BATTLE OF VERDUN

February 21-December 18, 1916

"They shall not pass."("Ils ne passeront pas.")
—General Robert Nivelle

SUMMARY: The Battle of Verdun lasted 300 days, cost 300,000 lives and 400,000 wounded men, all within the space of 10 square km—and no land gained. It is considered one of history's most horrific battles, and was essentially a war within the Great War.

At 0715 on February 21, after a 7-day delay because of weather, 1,400 German cannons and mortars opened fire on the fortresses of Verdun. Approximately a million rounds were used in the 10-hour opening bombardment. The next day, the Germans gained 5 km, up to the town of Flabas. On February 24, the French XXX Corps fell back to its second line of defense and the XX Corps showed up just in time to bolster the defense. The Germans pressed on and took Beaumont, then attacked Fort Douaumont, the largest fort in the Verdun defense. It had been built to accommodate 480 men and four cannon turrets, but at the time of the battle, held 68 men and maintained one operational retractable cannon turret. At 1630 on February 25, 14 officers and 79 soldiers of the German 24th took Fort Douaumont. To slow the German advance, the French counterattacked at Douaumont. Their ranks included then Captain Charles de Gaulle, who was captured by the Germans. The counterattacks worked, blunted the German attack and allowed 90,000 additional Frenchmen to arrive at the site of battle.

By March 2, the Germans had been delayed, but had also advanced too fast. They had not allowed their artillery to move up and provide covering fire, and as a result they suffered a pounding at the hands of the French artillery. The Germans managed to take the town of Douaumont, but an entire regiment was almost eliminated in the attack. At that point, the attack slowed.

On May 22, French Commander Robert Nivelle unsuccessfully attacked Douaumont with 3 divisions and 300 supporting guns. On June 2, the Germans counterattacked with 10,000 men and captured the top levels of Fort Vaux. For 5 more days, the Germans slugged it out door-to-door in the caverns of Fort Vaux until the French garrison surrendered because it had no water or food left.

The battle's next major phase arrived in July, when 60,000 Germans took Thiemont and obliterated the town of Fleury. Their next obstacle in getting to Verdun was Fort Souville. On July 10, the Germans used 60,000 shells of diphosgene gas ("Green Cross Gas"), although with little effect, because the French had recently upgraded their gas masks. On July 12, the Germans unleashed a 300,000-shell barrage on Fort Souville and advanced through the machine-gun fire and artillery. Only about 100 Germans made it to the top of the fort and caught a glimpse of Verdun from above before they were forced to retreat under the intense French artillery pouring down on them. That was considered the "high point" of the French defense and the farthest the Germans advanced.

On July 1, the Somme offensive began, helping to relieve the attack on Verdun and forcing the Germans to move men and artillery in an attempt to defend their northern sector from a breakthrough. At that point, the Germans replaced Commander Erich von Falkenhayn with Paul von Hindenburg.

The final major event at Verdun was the French attack on Fort Douaumont in October. For 6 days, the French pounded the fort with 530,000 75mm shells and 100,000 155mm shells. To thwart them, the French used a "creeping" artillery barrage that advanced ahead of the soldiers. They also brought to bear the 400mm rail guns located 13 km away, and six rounds were said to have penetrated the fort. On October 24, with the loss of 100,000 men, the French retook the remains of Fort Douaumont. Shortly after, on November 2, the French rolled up Fort Vaux and the Germans retreated. Finally by December 15, the French had pushed the Germans back to the lines they had maintained in February, 10 months earlier.

OUTCOME: The result was devastating losses in men and equipment on both sides, all for no ground gained. At Verdun, the French trucks, with their flexibility and speed, were said to have beaten the German logistic rail lines. Because of the trucks, the French surprised the Germans with their quick response of sending in reserves. The Germans were close to a breakthrough toward the middle of June, but when the British attacked in the Somme, Falkenhayn was forced to stop sending munitions and replacement divisions to the Verdun area. It proved to be a huge victory for the French, one from which the Germans never fully recovered. Falkenhayn's failure opened the door for Hindenburg and Erich Ludendorff to take control of the German part of the war. Also, as a result of the "success" of the Verdun forts, it heavily influenced French defensive doctrine and the French decided to build the Maginot Line after WW1.

WHO? The French commanders were Phillipe Pétain and Nivelle, who initially led only 30,000 soldiers. They faced 150,000 men commanded by Falkenhayn and Crown Prince Wilhelm, head of the Fifth Army. Over an area of 10 square km and spanning 300 days, the French suffered 400,000 casualties (including 163,000 deaths). The Germans suffered more than 300,000 casualties, with 140,000 killed. To this day, people notice a lack of birds in the area; legend says that so much death and destruction took place in Verdun, birds no longer choose to live there.

WHEN? February 21-December 18, 1916

WHY AT VERDUN? Verdun had 18 major forts surrounding it. After Liege and Namur in Belgium fell to German siege cannons, the French concluded that the Verdun forts were not worth fully arming. As a result, throughout 1915 the French streamlined their garrisons, stocking them with only 300 guns, limited ammunition, and a minimal number of men. The Germans saw the reduction of men and arms as an opportunity to take the "heart of France." Verdun was an imposing obstacle for the Germans, and since the age of the Romans had been home to a major fort guarding the northern approach to the fields of Champagne. A year earlier, the Germans had been unable to break through at Ypres; now they saw an opportunity to attack a location from which the French would not retreat. In addition, Verdun's capture would open the road to Paris. The Germans believed the French could not sustain their defense, because only a single road existed for logistical support. Verdun was on its own.

TECHNOLOGY/TACTICS: The Germans adjusted their tactics and used stormtroopers (stosstruppen) for the first time. In previous battles, they had relied on long artillery barrages followed by all-out troop assaults across wide fronts. The stormtroopers attacked after a short barrage of artillery and filtered through weak points, bypassed strong points to attack the rear area. Typically, they were heavily armed with grenades, machine pistols, regular pistols, and clubs. In addition, Verdun was the first battle during which the Germans used flamethrower teams to burn through strong points with devastating results. The French strongly believed in artillery as a deciding factor, hence French Gen. Philippe Pétain's quote, "firepower kills" ("le feu tue"). The French were correct, as artillery accounted for 70 percent of the casualties. The French fired more than 24 million rounds throughout the battle, the Germans more than 21 million.

Quote: "They shall not pass."(Ils ne passeront pas.) – General Nivelle

PLACES TO SEE: (These are some of the top sites, and many more exist.)

1. **MÉMORIAL DE VERDUN/FLEURY AND DOUAUMONT MUSEUM:** This is the area's best museum dedicated to the Battle of Verdun. It sits on the site of the Fleury train station.
   N 49 11.691 E 5 26.013 (49.19485, 5.43355)

2. **FLEURY VILLAGE CHAPEL AND 9 DESTROYED VILLAGES:** The town of Fleury once sat here. The town was at the center of battle and changed hands 16 times. It was one of 9 villages destroyed in the battle and never rebuilt. The other villages were Haumont, Beaumont, Louvremont, Ornes, Bezonvaux, Douaumont, Vaux, and Cumieres. Chapels were built in each of the destroyed villages to honor and memorialize the soldiers who died, and in many cases were buried there. It's a haunting experience to walk the paths, see the remaining foundations and realize that you can identify trenches and craters better than you can a town.

   A. FLEURY:
   N 49 11.866 E 5 25.689 (49.197767, 5.42815)

   B. HAUMONT:
   N 49 16.341 E 5 21.094 (49.272358, 5.351572)

   C. BEAUMONT:
   N 49 15.521 E 5 24.361 (49.258687, 5.406009)

D. LOUVREMONT:
N 49 14.285 E 5 23.918 (49.238082, 5.398638)

E. ORNES:
N 49 15.141 E 5 28.167 (49.252350, 5.469453)

F. BEZONVAUX:
N 49 14.201 E 5 28.080 (49.236684, 5.468001)

G. DOUAUMONT:
N 49 13.185 E 5 25.909 (49.219756, 5.431812)

H. VAUX:
N 49 12.348 E 5 28.204 (49.205793, 5.470068)

I. CUMIERES:
N 49 13.888 E 5 16.828 (49.231465, 5.280465)

3. **DOUAUMONT OSSUARY AND NATIONAL CEMETERY:** This is a powerful memorial and cemetery that houses a small museum. It is the burial place of 15,000 French soldiers, and also holds the bones (which you can see through ground-level windows) of more than an estimated 130,000 French and German soldiers who fought in the battle. The ossuary is a must-see when visiting Verdun.
N 49 12.479 E 5 25.438 (49.207983, 5.423967)

4. **BAYONET TRENCH:** This memorial honors French soldiers of the 137th Infantry Regiment who were buried alive by German artillery on June 23, 1916. Legend says several of the soldiers stood at the ready with their bayonets up. Hence, the memorial features rifles with bayonets sticking out. Unfortunately, vandals have broken much of what remains of the rifles.
N 49 12.805 E 5 25.519 (49.213417, 5.425317)

5. **FORT DOUAUMONT:** The largest and highest fort in the area of Verdun, Douaumont offers a tour during which you can walk around and on top of the fort to see the field of fire, destruction, and armaments. Douaumont was the first major fort taken by the Germans without a fight. Like many forts, the French had left it unoccupied because they believed the German 420mm howitzers would destroy them, as the Germans had done to Belgian forts in 1914. On May 8, 1916, the Germans tried to heat coffee inside the fort. That created a fire storm when flame thrower fuel caught fire and ignited artillery and grenades; 679 German soldiers died in the incident. The dead were walled off in one of the fort casements and remain there to this day. On October 24, the French recaptured the fort with 3 divisions. The recapture of Fort Douaumont marked the beginning of the end of the Battle of Verdun.
N 49 12.984 E 5 26.394 (49.2164, 5.4399)

6. **FORT SOUVILLE:** Another impressive fort with a short tour. You can climb to the top and explore the battle's immensity and destruction.
N 49 11.327 E 5 26.526 (49.188783, 5.4421)

7. **FORT VAUX:** The third major fort to explore around Verdun.
N 49 11.985 E 5 28.190 (49.19975, 5.469833)

8. **DOWNTOWN VERDUN:** Here you'll find many monuments, fortification remains, a river walk, and nice cafés and restaurants. Parking coordinate:
N 49 09.821 E 5 23.177 (49.163683, 5.386283)

9. **COLONEL DRIANT'S COMMAND POST AND TOMB AT CAURES WOOD:** At this location Colonel Émile Driant and his 1,200 men held off the German main assault, delaying it for about 24 hours. Driant was killed while rescuing a soldier as his troops withdrew. Only 118 of his men survived.
N 49 16.430 E 5 24.276 (49.273833, 5.4046)

10. **MORT HOMME MONUMENT TO THE 69TH DIVISION:** This is a memorial to the defenders who fought against the major German attack on March 6. The Germans took the hill after a month of fighting.
N 49 13.687 E 5 15.067 (49.228117, 5.251117)

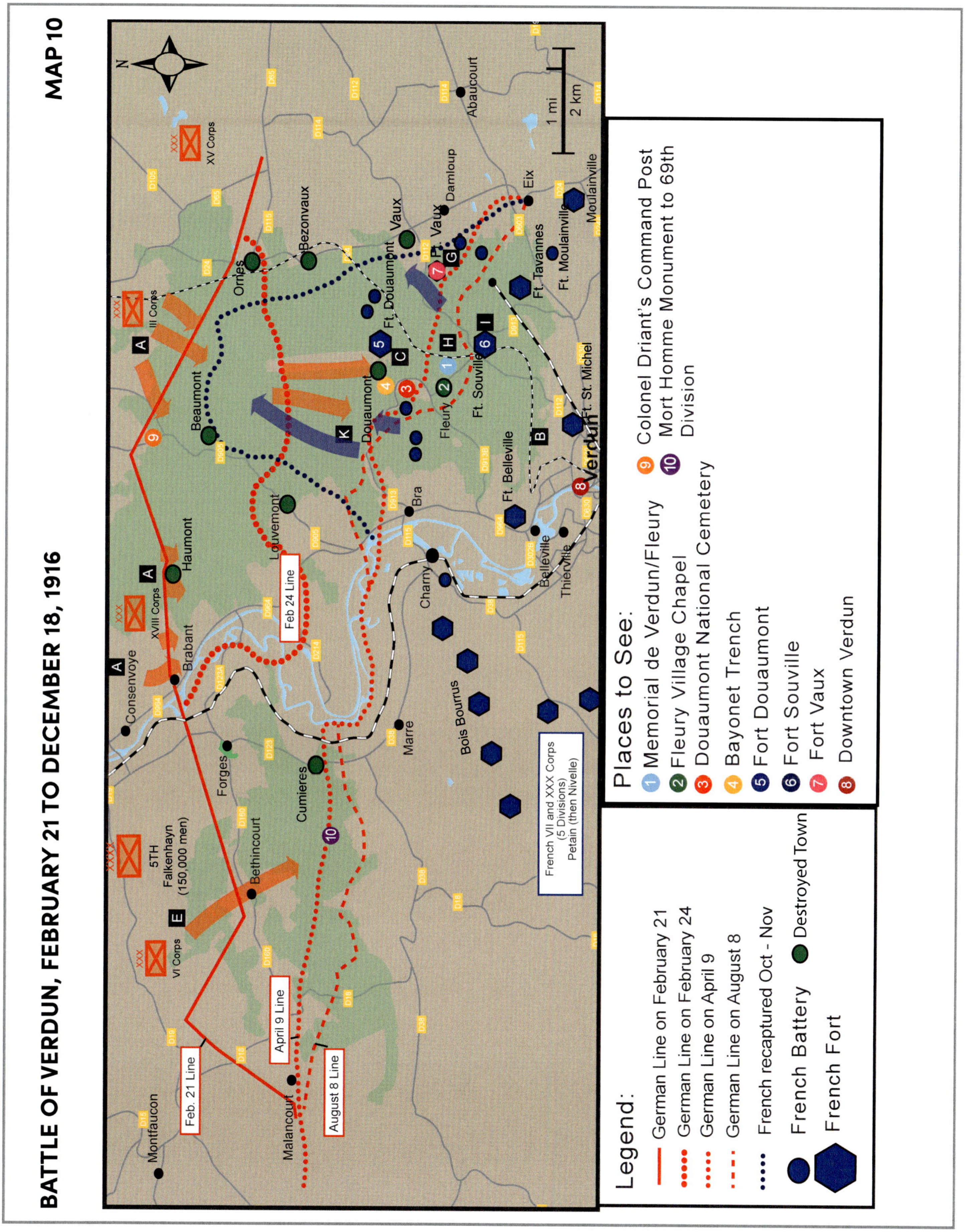
MAP 10
BATTLE OF VERDUN, FEBRUARY 21 TO DECEMBER 18, 1916
N
1 mi
2 km
XV Corps
III Corps
XVIII Corps
VI Corps
5TH
Falkenhayn
(150,000 men)
French VII and XXX Corps
(5 Divisions)
Petain (then Nivelle)
Feb. 21 Line
Feb 24 Line
April 9 Line
August 8 Line
A
B
C
E
G
H
I
K
Montfaucon
Malancourt
Bethincourt
Forges
Cumieres
Consenvoye
Brabant
Haumont
Beaumont
Louvemont
Ornes
Bezonvaux
Douaumont
Ft. Douaumont
Vaux
Ft. Vaux
Damloup
Abaucourt
Eix
Ft. Tavannes
Ft. Moulainville
Moulainville
Fleury
Ft. Souville
Ft. St. Michel
Ft. Belleville
Bras
Charny
Marre
Bois Bourrus
Belleville
Thierville
Verdun
Legend:
German Line on February 21
German Line on February 24
German Line on April 9
German Line on August 8
French recaptured Oct - Nov
French Battery
Destroyed Town
French Fort
Places to See:
1 Memorial de Verdun/Fleury
2 Fleury Village Chapel
3 Douaumont National Cemetery
4 Bayonet Trench
5 Fort Douaumont
6 Fort Souville
7 Fort Vaux
8 Downtown Verdun
9 Colonel Driant's Command Post
10 Mort Homme Monument to 69th Division

BATTLE OF VERDUN, FEBRUARY 21 TO DECEMBER 18, 1916

MAP 10

MOVEMENTS

A. February 21 at 0715 1,400 German artillery pieces opened fire until 1645. Then at 1645, six German Divisions of the 5th Army advanced. By nightfall they had only captured the first line of the French defense.

B. February 24, the French XX Corps arrived to help bolster the defense.

C. February 25 the Germans took Fort Douaumont, the largest fort.

D. French counterattacked at the town of Douaumont, slowing the advance.

E. March 6, the Germans tried attacking in the west at Mort Homme.

F. May 22 the French counter attacked unsuccessfully.

G. June 7, Germans captured Vaux.

H. June 24-26 the Germans used phosgene gas shells and almost broke through to Verdun.

I. July 10, Germans tried one more time to break through at Fort Souville but only saw a glimpse of Verdun.

J. October - November French recaptured Fort Douaumont and Fort Vaux.

K. December 15, the French pushed the Germans back to the original lines. The battle resulted in 300,000 dead and 400,000 wounded.

To left are photos of one of the destroyed towns; this one pictured is Haumont. It was located on the front line northen ridge of the battle. All that remains are ruined bunkers, building foundations and craters.

Battle of Verdun ended after 300 days with 300,000 dead and 400,000 wounded. Far left is a photo of a French soldier getting hit by fire. Left, the misery and reality of war. Many soldiers were buried in the trenches and just spilled out after the rain or artillery fire. (NARA)

Photo left is a shot of German soldiers observing an attack on Fort Vaux. (NARA) The photo right, Fort Vaux today.

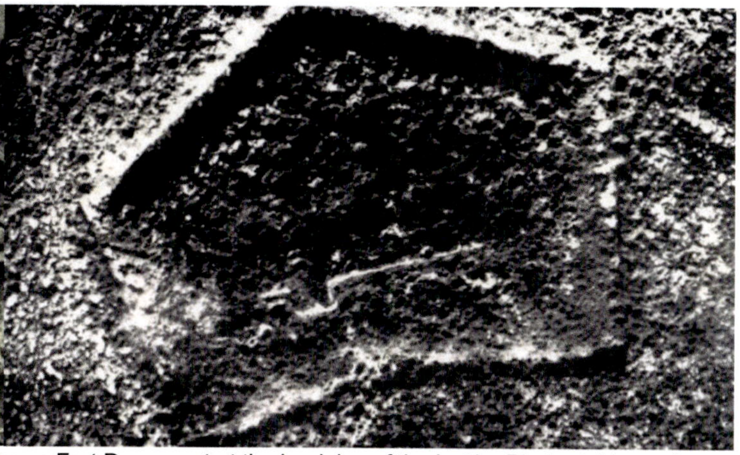

Upper left, shows Fort Douamont at the begining of the battle. Photo right speaks for itself, the same fort after the battle. The fort was the largest of the forts around Verdun.(NARA) Today lower left you can visit the fort, like Vaux to walk around and go inside. This photo is fro the top of the fort.

Verdun, downtown has many monuments including the beautiful walk way to the monument of victory flanked by two 75mm cannons. The photo to the right is a statue provided by Holland in memory of the French warriors, it depicts a winged Valkyrie carrying up a killed soldier to the heavens, to live forever.

Located in the memorial and Ossuary of Douaumont, they brought the human face to the war with photos of soldiers who survived the war. In the front of the tower are 15,000 French graves.Inside the Ossuary are the remains of 130,000 French and German soldiers found through-out the battlefield but not identified. Below left is the trench of bayonets, center the remains of Fleury and a chapel, right the Fleury before the battle.

BATTLE OF THE SOMME

July 1-November 21, 1916

"Good bye my loved ones. Don't cry."
—Last letter of Private John Scollen, 27th Northumberland

SUMMARY: On the first day of the Battle of the Somme, British forces suffered the greatest number of casualties ever recorded on a single day of fighting, a devastating 60,000. The battle continued for another 5 months. Similar to Verdun, it had enormous casualties and little movement in the lines.

In December, 1915, the British and French had drawn up plans to attack the Germans in the Somme region. The Germans upset those plans when they attacked Verdun in February, 1916, sapping any offensive capability the French had. As a result, the British forces took the lead in planning and providing the majority of the forces for the Somme. They initiated the attack on June 24 with an 8-day artillery bombardment, firing more than a million rounds at the German 1st, 2nd, and 3rd lines.

At 0720 on July 1, the British detonated a 40,000-pound mine under German positions at Hawthorne Ridge, followed by detonating 16 more mines at 0728. The Allies followed the massive explosions with an attack on a 30-km front with 280,000 men (13 British and 11 French divisions) against 10 German divisions. Their goal was the town of Bapaume, 19 km from their starting point of Albert. The old Roman Road was the axis point of the attack. The British command believed that after the bombardment they could "walk" to the German trenches, but their soldiers were slaughtered instead. For example, of the Newfoundland Regiment's 801 men, 500 instantly became casualties. The Irish of the 34th Division were almost wiped out less than 2 kms from the German lines.

The Germans had adapted their tactics to a "defense in depth." Since 1914, they had been largely unmolested, a fact that gave them time to create strong trenches, bunkers for machine guns and artillery. In addition, the British artillery did not pave a path through the German barbed wire as the British commanders assumed it would. The British artillery did little besides warn the Germans of an upcoming attack. In addition, the massive mine explosions were effective on the German 1st lines, but the German "defense in depth" prevented the British from advancing farther.

The French to the south, on the other hand, possessed more-accurate artillery and adjusted their infantry tactics so they were more like those of the German stormtroopers, creating small, high-firepower groups that effectively exploited weak spots in the opponent's defenses. The French achieved all their 1st-day objectives in the south, while the British suffered the worst losses in history. Of the 60,000 British casualties that day, 19,240 were dead, 35,493 were wounded, 2,152 were missing, and 500 were taken prisoner. The British met none of their objectives. To add to the horror, most of the casualties came from the same communities, because a man's hometown determined the unit in which he served. The losses devastated entire towns and villages.

On July 14, the British adjusted their tactics and implemented a short creeping barrage, attacking at 0325 with 4 divisions on a 6,000-yard front and achieving their objectives at Bazentin Ridge. Next, the 4th Army tried to take Pozières 3 times, but failed. The Anzacs (Australians and New Zealanders) took it on August 8 and held off German counterattacks. That was the Anzac's high point; like the Canadians at Vimy Ridge, they "became a nation on that day." The British took Guillemont on September 3 and the Irish 16th Division took Ginchy on September 9. Finally, the French and British linked forces again in a straight line, but at a tremendous cost to the British sector—1,000 yards gained at a cost of 82,000 casualties.

On September 15, the next major phase of the battle began at Flers-Courcelette. This is remembered as the first time tanks were used in battle, the new secret weapon of the British. Forty-nine tanks were delivered, and 32 made it to the line while only 21 made it into battle outside Pozières. The tanks were notoriously mechanically unreliable, but did have some "shock and awe" effect on the German lines. With 4 divisions and the tanks' help, the British captured Flers, and then High Wood.

On September 26, the final phase began. Field Marshal Douglas Haig convinced himself that a breakthrough was imminent. Marshal Ferdinand Foch wanted the British to keep up the pressure because the French were on the offensive in Verdun. Consequently, Haig threw in Hubert Gough's reserve army and on the first day, the 18th Division gallantly took the German-held fortress, Thiepval. Next, Haig moved forward with an attack north of Thiepval (later called the Battle of Ancre Heights), where Scottish Highlanders took Beaumont Hamel. Then the "Glasgow Boys Brigade" took the Frankfurt trench on November 18, but ended up surrendering 3 days later with only 45 survivors. That marked the end of the Battle of the Somme, the British losing 420,000 men for a gain of 3 km.

OUTCOME: The result was inconclusive. The Germans withdrew slightly to the stronger positions, but historians argue that without Somme, the Germans would have remained at full strength. The Battle of the Somme cost the Germans many veterans whom they would never recover, but it also did the same to the Allies. The British learned they needed to change their tactics, and paid dearly for that education.

WHO? The British forces comprising 13 divisions were commanded by General Haig and the French, commanded by General Foch, had 11 divisions. The two sides outnumbered the German forces led by commanders Max von Gallwitz and Fritz von Below, who together initially controlled only 11 divisions. The Allies suffered 623,907 casualties, of which 146,431 were killed. They also lost 100 tanks and 782 aircraft. The Germans suffered 595,294 casualties, with 164,055 killed in action. On average, the Allies suffered approximately 1,000 fatalities per day.

WHEN? July 1, 1916 - November 21, 1916

WHY AT THE SOMME? The British favored offenses in the Somme region because it was close to their supply routes via the English Channel. They planned to cut off the German U-boat threats supported from the Belgian coast. In addition, they wanted to relieve the pressure on Verdun.

TECHNOLOGY/TACTICS: To repulse the British forces, the Germans used defense in-depth, many lines, fortified strongpoints, and mobile reserves. The British, in turn, had developed tanks and introduced them to battle. Tanks were impervious to rifles and most of that era's machine guns, but were unreliable, slow (5 km per hour), hot, and susceptible to artillery. The tankers started wearing chain mail masks to protect their faces from spalling, i.e., fragments of steel shooting off inside the tank because of impacts on the outside. The French were more successful because they had adjusted their tactics to resemble those of the German stormtroopers; consequently, they had small bands of heavily armed troops to exploit weak points, bypass strong points and attack rear areas.

A German officer, Friedrich Steinbrecher, stated, "Somme. The whole history of the world cannot contain a more ghastly word."

Alan Seeger: One of the first Americans killed in action, Seeger lived in France and volunteered to fight for the country he loved, joining the French Foreign Legion. He wrote the following poem shortly before he died during the Battle of the Somme on July 4, 1916:

Rendezvous with Death

I have a rendezvous with Death
At some disputed barricade,
When Spring comes back with rustling shade
And apple-blossoms fill the air-
I have a rendezvous with Death
When Spring brings back blue days and fair.

It may be he shall take my hand
And lead me into his dark land
And close my eyes and quench my breath—
It may be I shall pass him still.
I have a rendezvous with Death
On some scarred slope of battered hill,
When Spring comes round again this year
And the first meadow-flowers appear.

God knows 'twere better to be deep
Pillowed in silk and scented down,
Where love throbs out in blissful sleep,
Pulse nigh to pulse, and breath to breath,
Where hushed awakenings are dear...

But I've a rendezvous with Death
At midnight in some flaming town,
When Spring trips north again this year,
And I to my pledged word am true,
I shall not fail that rendezvous.

PLACES TO SEE:

The area contains many memorials and cemeteries. A good starting point is the museum in Albert. Following are the highlights, listed from north to south:

1. **FRANKFURT TRENCH CEMETERY:** This is where the Glasgow Boys Brigade held out from November 18-21, 1916; only 45 survived. The Glasgow defeat marked the end of the Somme Offensive. One hundred and fifty soldiers are buried in this cemetery.
N 50 5.289 E 2 39.893 (50.08815, 2.664883)

2. **BRITISH MUNICH TRENCH CEMETERY (NEAR FRANKFURT):** At this location, approximately 20 Commonwealth soldiers are buried. The Munich Trench was near this site; the British captured it in 1917.
N 50 05.261 E 2 40.024 (50.087683, 2.667067)

3. **BEAUMONT-HAMEL NEWFOUNDLAND MEMORIAL:** This is another "must-see." It's a Canadian monument with preserved trenches, craters, barbed wire, and a welcome center. The location was part of the initial attack of July 1, 1916. Of the 801 soldiers who attacked, only 68 emerged unscathed. Note that at the time of battle, Newfoundland did not consider itself part of Canada. The soldiers' families bought the battlefield so they could preserve the hallowed ground. The war lasted another 2 years after the battle, so many of the remains were never recovered. The site also has a memorial to the Scottish Highlanders who captured Beaumont Hamel in October, 1916. The distance gained and the casualties suffered says much for the courage of the soldier, yet the futility of the attack.
N 50 04.430 E 2 38.917 (50.073833, 2.648617)

4. **ULSTER TOWER, THIEPVAL:** This is a monument to the Northern Irish 36th Division, which charged the German line from Thiepval Wood. They were the only soldiers north of the Bapaume road to pierce the German lines on the first day. Within 2 hours, they had penetrated five German lines, but after about 15 hours, they had to retreat to their original positions because no one on their flanks had gained ground.
N 50 03.658 E 2 40.799 (50.060967, 2.679983)

5. **THIEPVAL MONUMENT FRANCO, BRITISH MEMORIAL AND VISITORS CENTER:** This is the largest British memorial in the world. It contains the names of more than 72,000 missing troops who have no grave. This area, a high ground and German stronghold, was captured on September 27, 1916, but was lost in the 1918 Spring Offensive. The same day it was captured, Lt. Tom Adlam took actions that earned him the Victoria Cross. His citation reads: "A portion of a vil-

lage which had defied capture on the previous day had to be captured at all costs to permit subsequent operations to develop. This minor operation came under very heavy machine gun and rifle fire. Second Lieutenant Adlam realizing that time was all important, rushed from shell hole to shell hole under heavy fire collecting men for a sudden rush, and for this purpose also collected many enemy grenades. At this stage he was wounded in the leg, but nevertheless he was able to out throw the enemy and then seizing his opportunity, and in spite of his wound, he led a rush, captured the position and killed the occupants. Throughout the day he continued to lead his men in bombing attacks. On the following day he again displayed courage of the highest order, and though again wounded and unable to throw bombs, he continued to lead his men. His magnificent example and valor, coupled with the skillful handling of the situation, produced far-reaching results."
N 50 03.026 E 2 41.237 (50.050433, 2.687283)

6. **GUEUDECOURT (NEWFOUNDLAND) MEMORIAL:** This marks the farthest point taken in the Somme Offensive. It was captured by the Newfoundlanders on October 12, 1916.
N 50 03.901 E 2 51.229 (50.065017, 2.853817)

7. **POZIÈRES MEMORIAL:** This memorial where a windmill once stood marks the center of the struggle for Pozières. In August, 1916, 3 Australian divisions were thrown against the Germans to capture the Pozières high ground. In the course of 19 attacks over 45 days, they lost more than 23,000 men. Across the street is a monument commemorating the first use of the tank, which took place a month later. Also as of 2018, the memorial contains a memorial to the animals that fought during the war. While here, you can see Thiepval Memorial, not too far away. The town itself boasts 1 or 2 restaurants and a B&B. It is hard to imagine 3 divisions, almost 60,000 men, attacking this ground for more than 2 months.
N 50 02.698 E 2 44.168 (50.044965, 2.736130)

8. **TANK MONUMENT:** This site marks the first use of the tank, in September, 1916. The tank was 1st used during the attack of Flers-Courcelette, September 15-22, 1916. The memorial is an obelisk with small cast sculptures of tank designs. I hope one day we will be able to get a real Mk I tank there, or a replica!
N 50 2.650 E 2 44.198 (50.044167, 2.736633)

9a. TOMMY'S PUB, POZIÈRES: The food here is okay pub grub. Even more interesting is that the pub contains a great collection of photos and artifacts.
N 50 02.305 E 2 43.486 (50.038417, 2.724767)

9b. BRITISH CEMETERY, POZIÈRES: At this location on the western edge of Pozières is another Commonwealth cemetery. It holds the graves of 1,378 soldiers and commemorates the names of more than 14,000 soldiers who have no known grave.
N 50 02.028 E 2 42.927 (50.033796, 2.715444)

10. **MOUQUET FARM (CENTRAL BASTION IN THE GERMAN LINE ON POZIÈRES RIDGE):** At this site is a stone with a plaque describing the battle. It was the center of the British 2nd objective (after Thiepval).
N 50 02.968 E 2 42.483 (50.049467, 2.70805)

11. **DELVILLE WOOD SOUTH AFRICAN MEMORIAL AND MUSEUM, LONGUEVAL:** The South African Brigade attached to the 9th Scottish Division attacked the woods at dawn on July 15. After days of continual fighting and artillery strikes (more than 300 rounds per minute), only 143 men of the 3,150 originally in the brigade left the battle line unscathed. Across the street is the Commonwealth Cemetery of Delville Woods.
N 50 01.467 E 2 48.760 (50.02445, 2.812667)

12. **MAMETZ WOOD MEMORIAL (WELSH 38TH DIVISION RED DRAGON):** At this location, 2 divisions attacked Mametz Wood, one of them the 38th Welsh. They were repulsed, but a 2nd attack featuring bitter, 2-day, hand-to-hand combat forced the Germans to retreat. The attack left the Welsh division with 4,000 casualties; some blame them for the first offensive's failure, as they took too long to capture the stronghold.
N 50 00.830 E 2 45.381 (50.013833, 2.75635)

13. **PIPER'S MEMORIAL:** This is a statue dedicated to the pipers who played during attacks in the Somme.
N 50 01.572 E 2 48.209 (50.0262, 2.803483)

14. **LE GRAND MINE OF LOCHNAGAR:** At 0728 on July 1, 1916, this and 16 other mines exploded (in Le Grand Mine's case, thanks to 60,000 pounds of ammonal). Debris was thrown 4,000 feet into the air, but the Allies couldn't take advantage of the gap it opened up. The British soldiers took the German line, but soon after the shock, the German machine guns in the 2nd line of defense got back into the action, stopping cold the British attack. This is another must-see memorial.
N 50 00.933 E 2 41.833 (50.01555, 2.697217)

15. **15. MUSÉE DES ABRIS SOMME 1916, ALBERT:** This is one of the best museums dedicated to WW1. It's located below the church and a must-see. You should start your trip with a visit here.
N 50 00.252 E 2 38.918 (50.0042, 2.648633)

16. **COURCELETTE, CANADIAN MEMORIAL:** At this location is a memorial to the Canadians who fought during the Battle of Flers-Courcelette, September 15-22, 1916, during the overall Somme Offensive. The Allies gained more than 2,000 meters in some areas, but strategically it was still not the breakthrough they were looking for. The Allies suffered more than 29,000 casualties during the battle. Pvt. John Kerr, a Canadian soldier, was awarded the Victoria Cross for actions during the battle. The citation reads: "On 16 September 1916 at Courcelette, France, during a bombing attack, Private Kerr was acting as bayonet man and noting that bombs were running short, he ran along the parados under heavy fire until he was in close contact with the enemy when he opened fire at point-blank range,

inflicting heavy losses. The enemy, thinking that they were surrounded, surrendered. Sixty-two prisoners were taken and 250 yards of enemy trench captured. Earlier, Kerr's fingers had been blown off, but he did not have his wound dressed until he and 2 other men had escorted the prisoners back under fire and reported for duty."
N 50 03.228 E 2 45.120 (50.053794, 2.752007)

17. **BUTTE WARLENCOURT MEMORIAL:** At this location is an ancient mound that because of its position near the main road was important to capture. During the Somme, it was captured in October, 1916, and marks one of the farthest advances during the Battle of the Somme.
N 50 03.901 E 2 51.229 (N 50 03.901 E 2 51.229)

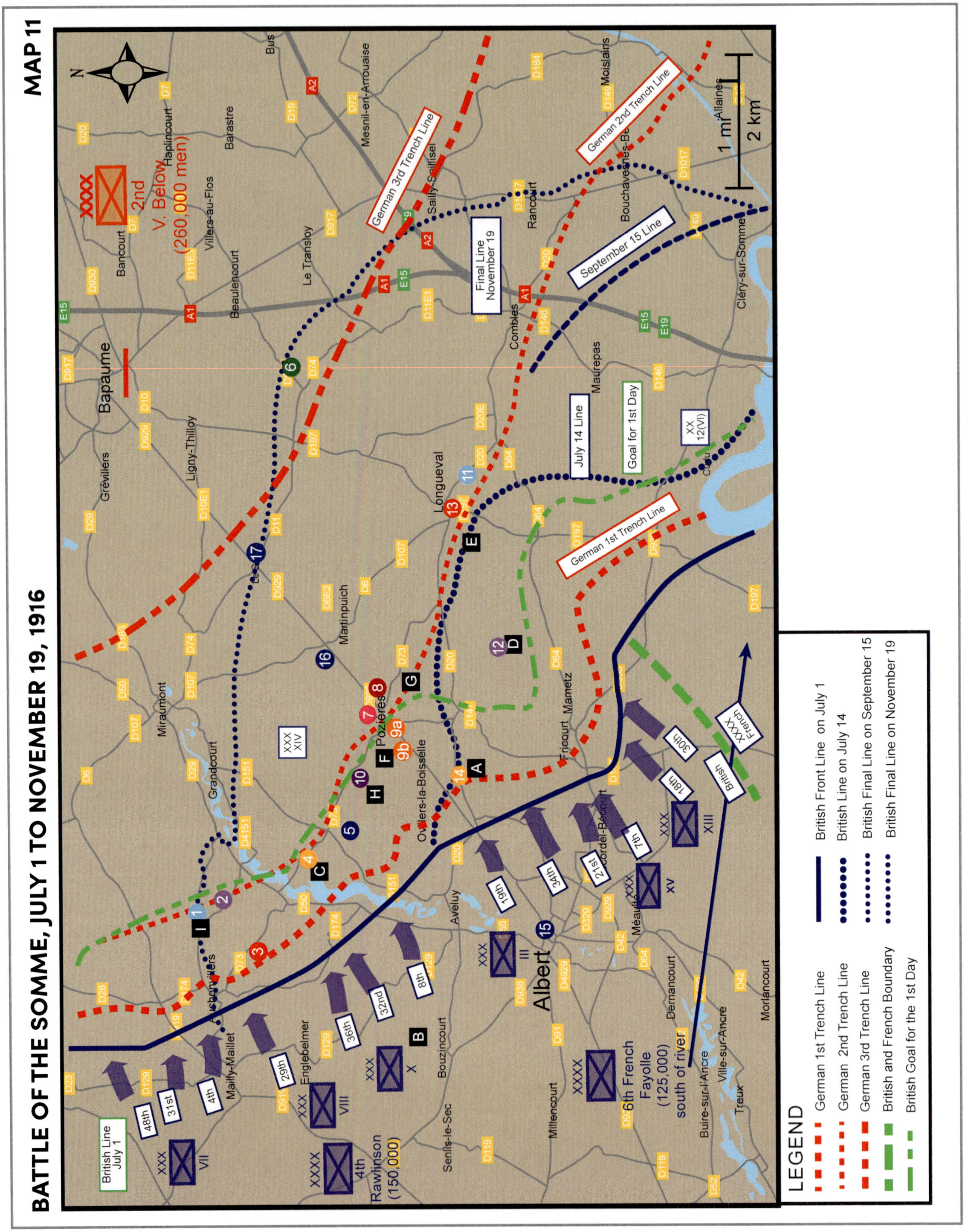

BATTLE OF THE SOMME, JULY 1 TO NOVEMBER 19, 1916
MAP 11
LEGEND
German 1st Trench Line
German 2nd Trench Line
German 3rd Trench Line
British and French Boundary
British Goal for the 1st Day
British Front Line on July 1
British Line on July 14
British Final Line on September 15
British Final Line on November 19
German 1st Trench Line
German 2nd Trench Line
German 3rd Trench Line
Final Line November 19
September 15 Line
July 14 Line
Goal for 1st Day
British Line July 1
2nd
V. Below
(260,000 men)
4th
Rawlinson
(150,000)
6th French
Fayolle
(125,000)
south of river
1 mi
2 km
Bapaume
Albert
Pozieres
Longueval
Combles
Maurepas
Martinpuich
Ovillers-la-Boisselle
Fricourt
Mametz
Miraumont
Grandcourt
Beaulencourt
Le Transloy
Sailly-Saillisel
Rancourt
Bouchavesnes-Bergen
Moislains
Allaines
Cléry-sur-Somme
Bancourt
Haplincourt
Barastre
Villers-au-Flos
Mesnil-en-Arrouaise
Grévillers
Ligny-Thilloy
Mailly-Maillet
Englebelmer
Bouzincourt
Senlis-le-Sec
Millencourt
Aveluy
Méaulte
Dernancourt
Buire-sur-l'Ancre
Ville-sur-Ancre
Treux
Morlancourt
Bécordel-Bécourt
48th
31st
4th
29th
36th
32nd
8th
19th
34th
21st
7th
18th
30th
VII
VIII
X
III
XV
XIII
XIV
British
French

BATTLE OF THE SOMME, JULY 1 TO NOVEMBER 19, 1916

MAP 11

MOVEMENTS

A. After an 8 day artillery barrage of 1,000,000 rounds; July 1 at 0728 the British detonated 17 mines, under the German Lines, including what is now known as the Lochnagar Crater.
B. At 0730 over 60,000 British soldiers attacked German lines and late morning another 40,000 were sent into the fray.
C. By the end of the first day the 36th Irish Division was the only one to breach the German line, in total the British suffered over 60,000 casualties in one day.
D. July 11, the Welsh finally captured Mametz woods after suffering 4,000 casualties.
E. July 14-15, the British and South Africans attacked at night with 20,000 men and gained ground around Longueval and Delville Wood.
F. August 4, the Austrialians captured the windmill at Pozieres but after 45 days of battle they suffered over 25,000 casualties for less than 2 miles.
G. September 15, the British secret weapon was introduced with success, 18 tanks were used in war for the first time and created a small breakthrough near Pozieres.
H. September 26-27, with the help of 3 tanks, Thiepval was finally captured.
I. November 13-18, one more final attack was attempted in the north, the 51st Highlanders caputured Beaumont Hamel and Frankfurt Trench but later surrendered marking the end of the Somme Offensive.

PLACES TO SEE:

1 Frankfurt Trench Cemetery
2 Munich Trench Cemetery
3 Beaumont-Hamel Newfoundland Memorial
4 Ulster Tower, Thiepval
5 Thiepval Monument Franco-British
6 Gueudecourt Memorial
7 Pozieres Memorial
8 Tank Monument
9a Tommy's Pub
9b British Cemetery Pozieres
10 Mouquet Farm
11 Longueval-Delville Wood Memorial
12 Mametz Wood Memorial
13 Piper's Memorial
14 Le Grand Mine of Lochnagar
15 Musee des Abris Somme 1916, Albert
16 Courcelette Memorial
17 Warlencourt Memorial

Left to Right: One of the connecting roads headed from the Frankfurt Trench towards Beaumont-Hamel. The poppies were coming out in full bloom and the bright yellow rapeseed seen to the left which is used as cooking oil, biodiesel and animal feed. In the center is the monument to the bag pipers and to the right, Delville Wood Cemetery and Memorial for the South Africans.

Photo left, a Germantrench, notice the better quality compared to the above right photo of the British trench which was typical of the Allies who thought that they were always going to be on the attack, so why build up a trench system they were going to leave? (NARA)

Right, the Ulster Tower to commemorate the 36th Irish Division who pierced the German line on the first day of the Somme but had to retreat after 15 hours because no one else was on their flanks.

Below left, the British Commonwealth Cemetery Pozieres at sunset. To the right, after the war, family from England traveled to France to find the graves of their loved ones, the terrible truth. (IWM)

The left side are clips from real film footage of British soldiers attacking during the Battle of the Somme.The second photo is gut wrenching to see the soldier look back at the camera for possibly one last time. It appears he was hit as he went over. The courage it took to attack is incredible. On the right are photos of the Germans in the attack. The Germans began using tactics to employ storm troopers, soldiers with grenades and submachine guns to bring more fire power. (NARA)

Photo upper left, Somme Museum in Albert is my suggested starting point. The center photo is a photo of one of the 16 detonations that took place at the beginning of the Somme.The photo is from a video of the Hawthorne Ridge explosion which used 30,000lbs of TNT.
The above photo is a memorial outside of Albert called, the Grand Mine of Lochnagar. It was blown on July 1, 1916 with 60,000 lbs of ammonal. You can barely make out the people in the photo on the far ridge. The photos to the left are of the Canadian Memorial Park, Beaumont-Hamel, Newfoundland. A gruesome photo of a German soldier's remains next to his defenses in that area. (NARA)

Upper left is the memorial outside Pozieres where 3 Austrialian IDs attacked 19 times over the land you see above in the black and white photo of Pozieres fields. On the bottom left is Mouquet Farm today which was a German stronghold, facing towards Pozieres.
To the right a British MK I tank, first used in the Somme near Pozieres.

Thiepval Memorial, in the distance, one of the British main targets to capture. Today, it is a memorial to over 72,000 soldiers with no known grave. Photo right is from the center of the memorial, looking north along the line.

Above Left: Gueudecourt Newfoundland Memorial, a monument to the furthest advance of the Battle of the Somme, which took place in October 1916. To the right, also another memorial to the north, marking one of the farthest advances, Butte Warlencourt, marking the furthest the British advanced during the Somme on the road to Bapaume

Photo left, is the Canadian Memorial Courcelette, which marks the area captured by the Canadians in action during the Battle of the Somme, September 1916.

BATTLE OF FROMELLES

July 19–20, 1916

"Don't forget me cobber!"
—wounded soldier to Sgt. Frazer

SUMMARY: The Battle of Fromelles was a total disaster that nearly destroyed the Australian forces involved. The British launched their major offensive of the Somme on July 1, 1916. The Battle of the Somme, which was going extremely poorly, was in response to the French being battered in the Battle of Verdun. The Fromelles battle was devised to take advantage of the believed-to-be weakened German line north of the Somme and assist in diverting German forces. The British command initially wanted to attack on July 15, but delayed the attack because of weather and other preparations. The target for the operation was the German strongpoint of Sugarloaf Salient, part of the high ground of Aubers Ridge, which the Germans held.

The British ordered the XI Corps of the 1st Army to conduct the operation. The Corps contained the weak British 61st ID and the rookie 5th Australian ID. The Australian division was about to get its first experience in a major operation. Facing the Allies was the German 6th Bavarian Reserve Division, flanked by other divisions in support. The German reserve division was made up of a scramble of poorly trained, very young and old troops, but the Germans were provided additional machine guns and bunkers. They also held the high ground overlooking the battlefield.

The battlefield was on flat terrain, cut by various small canals and streams. The ground was marshy in places and there was no use in digging deep trenches, so both sides stuck with building small battlements of dirt and sandbags. The Germans built stronger concrete structures.

On the day of the battle, the bombardment began at approximately 1000 and lasted until 1730 when the infantry began its assault. The Australian 5th Division was on the north and the 61st ID on their right flank. The 5th deployed its brigades from north to south, the 8th Brigade on the left flank (north), 14th in the center, the 15th on the right flank. The daylight attack was immediately met by German high-explosive artillery and machine gun fire. On the right, the 61st ID units were quickly bogged down and canceled their attack without getting word to the Australians. Despite severe casualties, the 8th and 14th brigades made some progress and dug in. The 15th Brigade attempted to attack Sugarloaf 4 times, and at 2120, the 15th reserve units attacked Sugarloaf, but both units were decimated.

At 2300, the German 21st infantry regiment counterattacked the Australian 14th Brigade, inflicting more casualties. Then at 0200, the Germans counterattacked in greater force, cutting the Australians off from one another in many places. At 0330, the Allied command ordered a withdrawal, but it was too late. The majority of the cut-off Australians fought for 6 hours to get back. During that time, the Germans agreed to a truce that allowed some of the Australians to return to their lines. The next day, the planned July 20 attack was canceled. The Australians lost more than 5,000 men in under 24 hours. It is considered the worst day in Australian military history.

OUTCOME: It was a total disaster for the Allies. Because of the disaster of Gallipoli and of the Somme, however, the reporting of it was minimized to avoid upsetting the public more than it already was, especially the Australians. The Germans held their own and did not have to move troops or reserves at all, thus not experiencing a negative impact on their Somme defense.

WHO? The British attacked with 2 divisions, approximately 30,000 men, the 61st and Australian 5th IDs against the Bavarian 6th Reserve Division with approximately 12,000 men. In less than 24 hours, the Allies suffered 7,080 casualties, of which 5,533 were from the Australian 5th ID. The Germans suffered about 1,200 casualties.

WHEN? July 19-20, 1916

WHY AT THIS LOCATION? The British believed this was a weak point and wanted to divert German reserves to this sector from the Battle of the Somme.

PLACES TO SEE:

1. **FROMELLES MUSEUM, "ASSOCIATION POUR LE SOUVENIR DE LA BATAILLE DE FROMELLES":** This museum is well done and located on the 2nd floor of the town hall. Just be aware of opening times.
   N 50 36.349 E 2 51.226 (50.605817, 2.853767)

2. **FROMELLES MEMORIAL PARK:** At this location, a farmer donated his land to the Australians who fought there during the Battle of Fromelles, July 19-20, 1916. Also here is the famous "Cobbers" statue that depicts Sgt. Frazer rescuing an unknown comrade. The park is located on what was the center of the battlefield and German lines, and contains a few bunkers, map, and a good view of the Sugarloaf objective.
   N 50 37.078 E 2 50.130 (50.617967, 2.8355)

3. **VICTORIA CROSS AUSTRALIAN CEMETERY AND MEMORIAL:** This cemetery is unique for two reasons: It is the only all-Australian cemetery on the Western Front and it does not have any headstones. Because of the battlefield, it has 2 mass graves with 410 casualties buried, so the casualties are memorialized by 410 rose bushes. It also commemorates 1,299 dead who have no known grave. It is located on what was no-man's land during the battle.
   N 50 37.166 E 2 50.012 (50.619437, 2.833538)

4. **SUGARLOAF REDOUBT LOCATION:** Although not easily accessible, this is geo coordinate is on a road close to the location of the German Sugarloaf Redoubt that both the British and Australians tried to overcome. It is from the perspective of the German lines, the redoubt was located approximately 200 meters west from this location.
N 50 37.166 E 2 50.012 (50.619437, 2.833538)

5. **LE TROU AID POST CEMETERY:** At this location, the British had set up the Australian Brigade HQ during the battle. It now contains the graves of 356 dead from the battlefields around this area. Like all the cemeteries, it is very beautiful and well done for remembering their sacrifices.
N 50 37.428 E 2 49.601 (50.623793, 2.826688)

6. **PHEASANT WOOD CEMETERY:** At this location in Fromelles is a Commonwealth Cemetery that contains the graves of 220 Australian and 70 British soldiers. It was built only recently and dedicated in 2010 after remains were found in a mass grave of soldiers buried by the Germans after the Battle of Fromelles. It is located next to the museum.
N 50 36.483 E 2 51.049 (50.608046, 2.850812)

7. **BUNKER DE L'ABBIETTE:** A German WW1 bunker in the center of the battle, one of the objectives of the Australians. Today, there is a park and you can walk to the bunker where there is a good view of the battlefield and an information board. At this coordinate is a farm road where you can park and walk in about 100 meters to the bunker.
N 50 36.630 E 2 50.658 (50.610505, 2.844300)

BATTLE OF FROMELLES, JULY 19–20, 1916

MAP 12

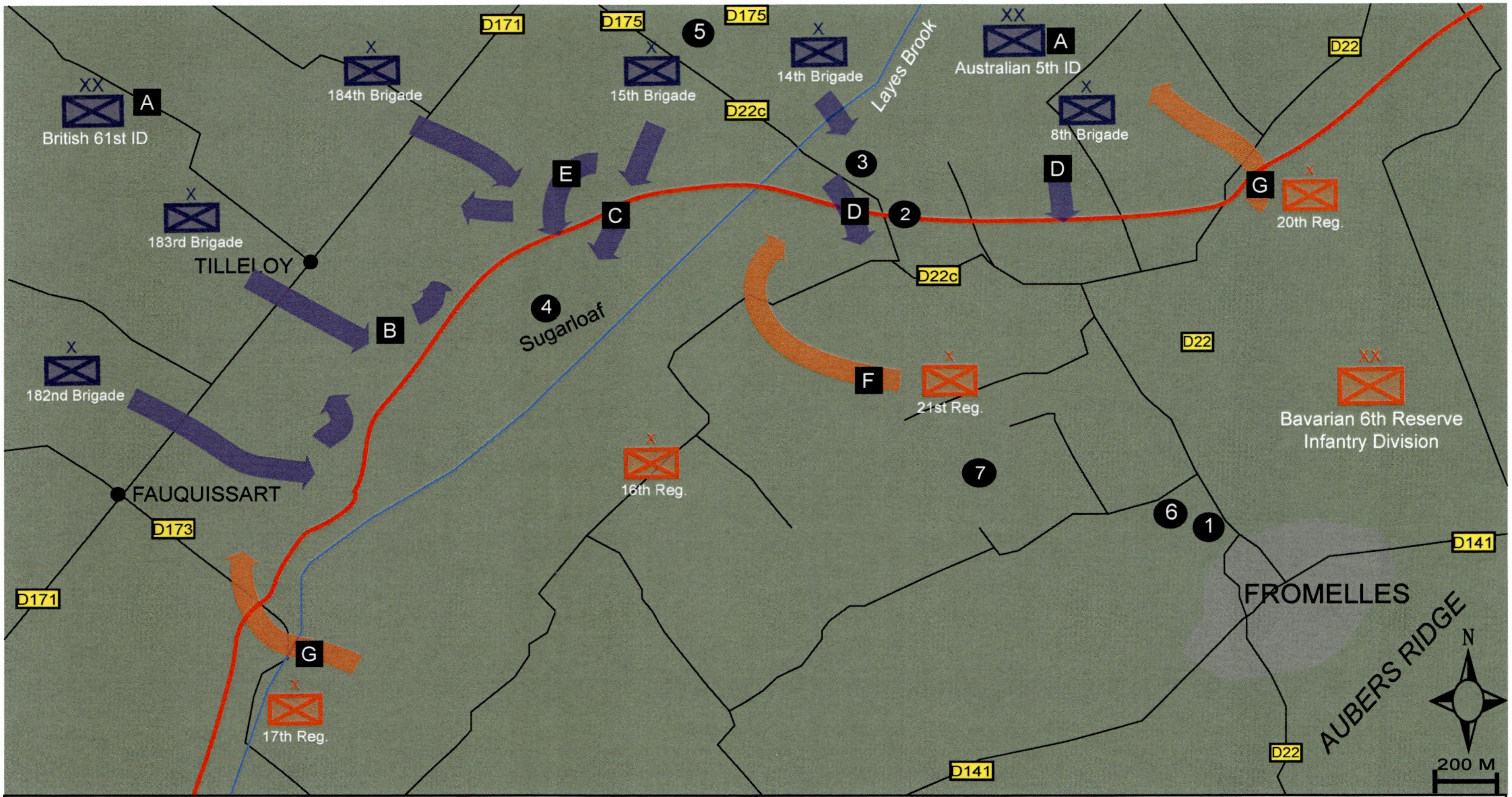

MOVEMENTS:
A. After a 7 hour artillery bombardment, at 1730 the infantry attacks.
B. The 61st ID is blunted by fierce German attacks.
C. The Australian 15th Brigade is devastated after attacking Sugarloaf 4 times.
D. The 14th and 8th Brigades made some progress and dig in.
E. The 15th Brigade reserves attacked Sugarloaf at 2120 but are also repulsed.
F. The German 21st Regiment attacked the the 14th in their flank at 2300.
G. The Germans attacked in greater force on the flanks at 0200.
H. The Allies called for a retreat at 0330.

PLACES TO SEE:
1. Fromelles Museum
2. Fromelles Australian Memorial Park
3. VC Australian Cemetery and Memorial
4. Sugarloaf Redoubt location
5. Le Trou Aid Post Cemetery and Brigade HQ
6. Pheasant Wood Cemetery
7. l'Abbiette Bunker

German Front Line

THE HINDENBURG LINE

1916 -1918

"Fortifications are monuments to men's stupidity."
—General George Patton

SUMMARY: After the 1916 ghastly losses in men and equipment during the Battles of Verdun and the Somme, the Germans struggled to maintain a strong Western Front position. To effect change while both battles were winding down in September, German Chief of Staff Eric von Falkenhayn was replaced by General Paul von Hindenburg. That same month, Hindenburg assembled the German command staff in Cambrai to update their strategy to a defensive posture, consolidate their forces and update their defensive doctrine. Rather quickly, they studied the feasibility of shortening their front lines and building new defensive fortifications. For example, between Arras and Laon, they could shorten their front line from 142 km to 45 km, freeing up 8 to 10 divisions, not to mention large amounts of artillery and ammunition. The Germans at that time also quickly adjusted their defensive doctrine to reduce unnecessary casualties by focusing on an elastic defense to destroy the enemy forces versus the old doctrine of holding territory.

After the conference in Cambrai, in October, 1916, the Germans decided on where to build their defenses and adjust their front lines. Across the entire Western Front, the Germans studied the terrain and assembled large amounts of materiel for the fortifications, such as concrete, rebar, barbed wire, barbed wire stakes, communication lines, and much more. The main defensive lines were called "Stellungs" or positions. They maximized the terrain, towns, and even canals to aid in the fortifications, obstacles, and to make the artillery and machine guns as deadly as possible. The main defensive lines were:

1. **WOTAN STELLUNG:** Located behind the lines at Vimy, Loos and Arras. It ran from Armentieres to Peronne, approximately 80 km. It did not shorten the line, but it pulled them back to better terrain. The Germans began construction on this line in January, 1917 and occupied it in April that same year.
2. **VERDUN SECTOR, MICHEL STELLUNG:** In this sector, they built a line at the base of a Salient that was 60 km long and freed up five Divisions. The position construction began in July, 1917 and was not occupied until September, 1918.
3. **FLANDERN STELLUNG:** In the much fought-over area of Ypres and low water line, the Germans built a line from the coast along the Passchendaele Ridge to the Wotan Line, 10 km behind the front lines, doubling the defensive depth. A higher-priority area, the Flandern position construction began in June, 1917 and was hastily occupied that August.

4. **HUNDING STELLUNG:** This position was built to contain the Noyon and Champagne battlefields. It was planned for 240 km, of which 215 km were built, running from Wotan near Peronne along the Somme River to La Fere to the north of Verdun. Construction for Hunding began in March, 1917, but was never occupied.
5. **SIEGFRIED STELLUNG:** As stated above, it ran from Arras to Laon and was designated the highest priority for materiel and labor. Construction began in September, 1916, and it was the first to be occupied, in March, 1917.
6. **OTHER POSITIONS WERE PLANNED,** such as in Alsace and Lorraine, but because of the terrain, other older forts, and low amounts of fighting, the Germans did not commit the resources to build them.

In February, 1917, the German command decided to withdraw from Arras to Laon sector, "Siegfried Stellung." They were crafty in maintaining a low signature that did not give away the withdrawal plan, called Operation Alberich. On March 16, the operation commenced, and in a brilliant act of organization and deception, the Germans withdrew 35 divisions and materiel to the new defensive position in fewer than 4 days. During their withdrawal, they razed towns, poisoned wells, and booby-trapped positions. They left about 30 percent of the divisions to protect the withdrawal just in case the Allies attacked, but the Allies were slow to react. It was a success and the Germans were able to free up 13 divisions and their supporting artillery.

The new defensive positions included use of new defensive doctrine to save lives, compared with the older doctrines. The pre-war doctrine focused on getting all forces on the front line to maximize fire power and stop the enemy. And if attacked, always gegenstosse! (counterattack), which the Allies eventually anticipated, decimating the counterattacking troops.

In 1915, after the initial devastating losses during the first year of the war, the Germans adjusted tactics to fortify their front line by having two lines about 150 meters apart, with bunkers, belts of barbed wire up to 30 meters deep, and artillery close by. They also included the tactic of "hinterhangstellung," holding the reverse slopes of hills to reduce Allied artillery effectiveness. In addition, the "1915 doctrine" called for a support-trench system approximately 4 km behind the front line, with the reserve troops ready to counterattack. Again, the focus was to hold the front line until the counterattack was assembled and relieve the front line to win back territory. That doctrine was better than the pre-1915 one, but it still resulted in terrible casualties, because the Allies knew the Germans always counterattacked, and were ready for it.

Next after the Somme and Verdun in the fall of 1916, the Germans developed a new doctrine based not on holding the ground, but being elastic and destroying attacking enemy forces. It used the Hindenburg Line to create a deep zone of defense and was flexible to allow the attacker into the zone, then destroy him with artillery and machine-gun fire. The new German positions zigzagged across the terrain, using canals, towns, and forests to their advantage. They organized the line with 3 zones. The 1st was the lightly manned outpost for an early warning, the 2nd the battle zone, and the 3rd the rearward with 50 percent of the forces in reserve or counterattacks. The 3 zones created a defensive line about 15 km deep. The zones included barbed wire, in some places 30 meters deep; irregular trenches; bunkers; troop shelters; gun emplacements; and observation posts. They used so much barbed wire, the German industry could not keep up with the requirements. The Germans continued to use the "hinterhangstellung" as well, which put the machine-gunners at a disadvantage for fields of fire, but it saved many lives. The trenches were typically 1 meter wide at the bottom, 3.5 meters wide at the top and 2 meters high/deep. The shelters, a "safe home" from artillery fire, could withstand direct hits from 150mm rounds and some of the rear-position shelters could withstand 210mm rounds. The Germans also included more fire-power to the units. They increased the trench mortars and machine guns in each company to 6 each.

The new defensive doctrine was first tested successfully during the Nivelle Offensive of April and May, 1917. Later, the Siegfried Line was tested in force during the November, 1917, Battle of Cambrai and it also held. Just prior to the Battle of Cambrai and confident in his forces, Hindenburg decided on November 11, 1917, for a 1918 Spring Offensive before the Americans would be fully engaged in the war.

His decision, exactly a year before the end of the war, required the Germans to leave their excellent defensive positions and suffer greater losses. After the failed Spring Offensives, they retreated to the Hindenburg Line positions across the front, in many cases ill-equipped to defend against the Allied onslaught who used new combined-arms tactics. From August, 1918, to November, 1918, the Allies punched through at different points of the weakened Hindenburg Line. The Germans could have withdrawn farther, but unable to, they collapsed and sued for peace.

OUTCOME: The line was successful for about a year, until the Germans went back on the offensive. Then, Allies began using combined arms against the line and were able to break through.

WHEN? September, 1916-November, 1918.

WHY IN THIS REGION? It was built on the Western Front to counter the higher amounts of Allied divisions versus the German ones.

PLACES TO SEE:

Throughout this book, there are a variety of places where one can see the bunkers and remains of the Hindenburg Line. Many times, the remains of the line are incorporated into a cemetery or memorial where the soldiers fought. Unfortunately, at this time there is no dedicated museum that I am aware of.

THE HINDENBURG LINE

MAP 13

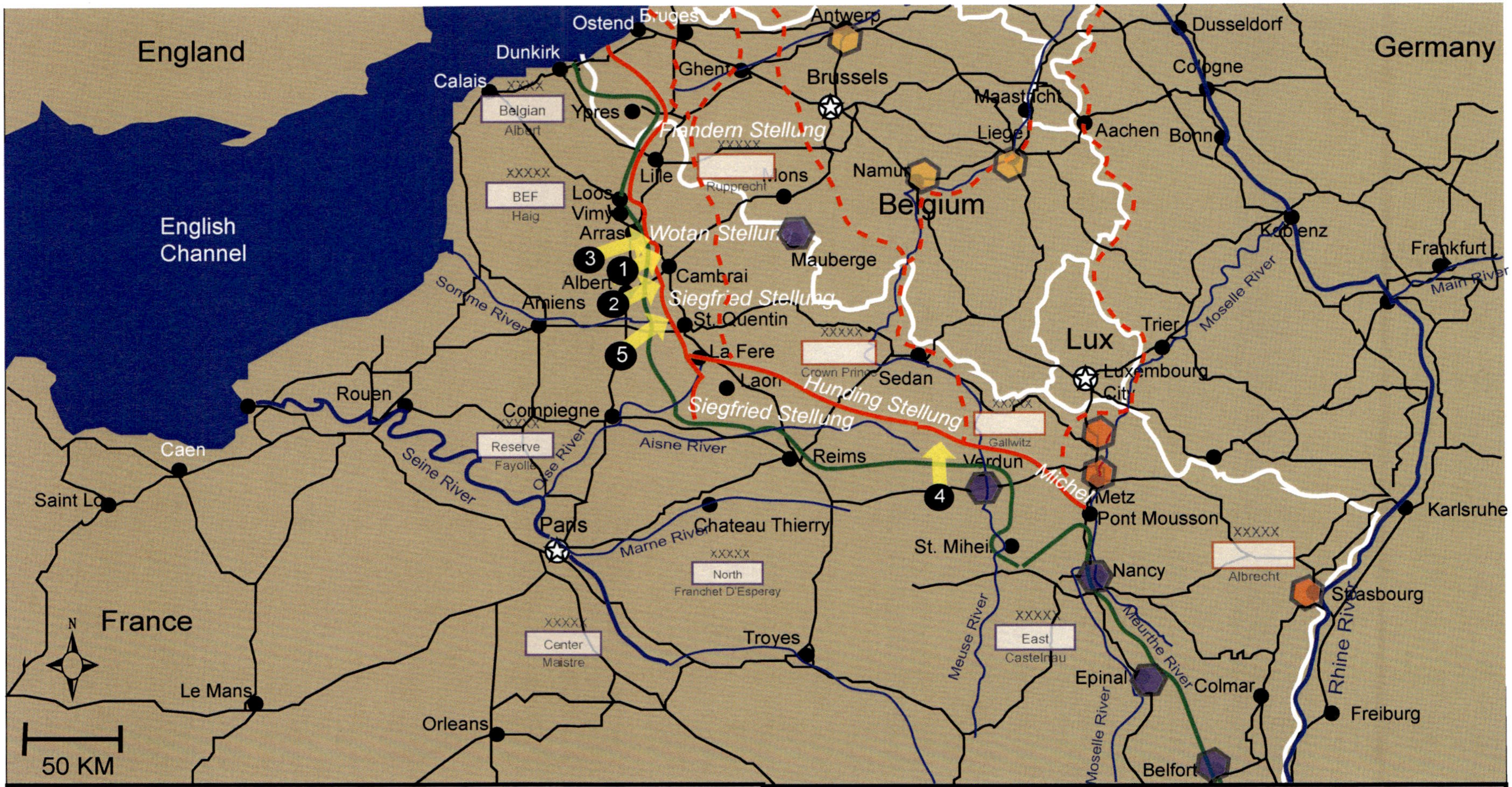

Above provides a rough view of the "Hindenburg Line" marked in red against the green front line of 1917. It also provides a quick view of the major thrusts into the Hindenburg Line by the Allies. The following are major opeations that challenged or broke through the line:

1. Battles of Bullecourt, April 11 and May 3-17, 1917
2. Battle of Cambrai, November 20 - Dec 7, 1917
3. Battle of Amiens, August 8-12, 1918
4. Meuse Argonne, September 26-Nov 11, 1918
5. Saint Quentin Canal, September 29-October 3, 1918

KEY:

- Situation Spring 1917
- French Fortress
- German Fortress
- Belgian Fortress
- Hindenburg Line & Stellung Names
- Planned Lines
- Major Allied Attacks on the Line

NIVELLE OFFENSIVE

Aisne, Champagne, April 16–May 16, 1917

Battle of Vimy Ridge, April 9–12, 1917

"In those few minutes I witnessed the birth of a nation."
—Canadian Commander Brigadier General A.E. Ross

SUMMARY: The Battle of Verdun officially ended in mid-December of 1916, and the new French commander, Robert Nivelle, assumed control of the Allied efforts. He drew up plans for the Allies' Spring Offensive, which became known as the Nivelle Offensive. His plan was to have the British and Commonwealth forces create a diversionary attack in the north at Arras and Vimy. Next, the French would attack with 1.2 million men and 7,000 cannons on an 80-km front, with the goal of making a major breakthrough on Chemin des Dames Ridge just west of Reims. Nivelle believed the maneuver would end the war within 48 hours and result in only 10,000 casualties. The result, however, was far different from Nivelle's predictions.

On March 20, the British and Canadians initiated an intense barrage on a 38-km front at Vimy Ridge east of Arras. They fired more than 2.68 million rounds, including gas shells, for the last 10 hours before zero hour on April 9. The British and the Commonwealth, which consisted of 27 divisions, were up against 7 German divisions on the line and 27 more in reserve. Unlike the Somme, the artillery bombardment at Vimy proved effective against the Germans, because it occurred over a long period, disrupting ration resupply, destroying barbed-wire defenses, and killing many pack animals with gas rounds, consequently ensuring that the Germans would encounter difficulty moving their artillery units.

On April 9 at 0530, in the dark and a heavy snow, the infantry attacked behind a wall of creeping artillery fire. Soon after, the Allies captured 2 German front lines. By 1300 on the first day at Vimy Ridge, Canada's 4 divisions, fighting together for the first time, took the majority of the ridge from 3 German divisions. The Canadians' success stemmed from extensive training, thorough planning, and specific objectives for each platoon, enabling them to continue fighting even if their officers were killed. In addition, the 4 previous months saw 55 trench raids that harassed the Germans and gathered intelligence. The Canadians were heavily armed with 80 machine guns for every brigade and a Lewis gun for every platoon. On the attack's 2nd day, they captured the town of Thélus. On the 3rd day, they captured Givenchy en Gohelle. On the German side, Commander Alexander von Falkenhausen was blamed after the battle for not applying the German doctrine of "elastic defense." As a result, the German forces were overwhelmed in static positions, with the reserve forces too far away to counterattack.

On April 20, shortly after the diversionary Battle of Vimy Ridge, the French opened up their offensive in the south. The battle started with a massive military bombardment, followed by an attack of infantry supported by tanks. Unfortunately, the Germans were ready for the French, because of gathered intelligence and French delays. The Germans prepared their defenses in-depth and placed 100 MG08 machine guns every 1,000 meters. The French initially used tanks and men to overwhelm the first German lines, but the Germans responded with pre-measured artillery and destroyed the French attack.

The French had a minimal gain of approximately 3.2 km and suffered a great amount of casualties. Just in the initial 5 days, the French had 120,000 casualties, 40,000 of them on the first day alone. They also lost 140 tanks. Nivelle still believed, however, that they could break through, and on May 5 launched a final assault along the German lines, with little gain. As a result, French citizens grew angry and soldiers mutinied. In one sector, elements of the French 2nd Division showed up drunk with no rifles. After 21 days of fighting, on May 9 the offensive that was supposed to end the war in 48 hours was itself over.

OUTCOME: Overall, the Nivelle Offensive was a failure, although the British and Canadian sectors achieved some gains, such as Vimy Ridge. After the success of Vimy Ridge, it was said that a "nation was born" because under their own command, Canada fought heroically, effectively, and emerged victorious.

As for the French, Nivelle was sacked as commander and replaced by a less aggressive General Phillipe Pétain. The failure of the battle led to widespread French mutinies that the French suppressed with executions. The Germans sacked Falkenhausen because he had not followed German defense doctrine in the Arras salient, thereby enabling the Canadian and British to capture territory.

WHO? The French had 1.2 million men under Nivelle, suffered 270,000 casualties, and lost 140 tanks. The Germans under Fritz von Below maintained a force of approximately 480,000 soldiers against the French and suffered about 140,000 casualties. Against the British and Canadians, the Germans had 400,000 men under Falkenhausen and suffered about 120,000 casualties. The British under Douglas Haig had about 350,000 men and suffered 147,656 casualties. The Canadians had a force of 170,000 men under Julian Byng and suffered 11,004 casualties (including 3,598 killed in action) while taking Vimy Ridge.

WHEN? Arras and Vimy Ridge took place April 9-12, 1917, and the Nivelle Offensive took place in Champagne from April 20-May 9, 1917.

WHY AT ARRAS AND CHAMPAGNE? The British attacked at Arras because it was close to their supply routes and an are ht defense in the first line and several hardened lines farther back. Artillery and reserves in the rear waited to provide support when needed. At this time, the Germans had started producing more than 14,000 machine guns a month, including the lighter version of the MG08. The Allies were using tanks, including the French Char Schneider tank, to little effect. Furthermore, the British used mines and tunnels, primarily built by New Zealanders, to move supplies, provide housing, and enable surprise attacks on the German lines. In terms of artillery, the

British used "creeping" barrages that fired at set distances to maintain a length of 100 meters ahead of the advancing forces (and thereby keeping the Germans in their bunkers instead of on their machine guns). The Allies found that their artillery was more successful in this battle because they could use "instantaneous fuses" (which exploded at the lightest touch) against barbed wire.

Song of Craonne (chorus)

Adieu life, adieu love
Adieu all the women
It's all over, it's for good
This ghastly war
At Craonne on the plateau
We had to lose our hides
Because we are condemned
We are the sacrificed.

PLACES TO SEE

THE NIVELLE OFFENSIVE IN CHAMPAGNE:

1. **NATIONAL MONUMENT TO TANKS:** Unfortunately, the monument uses post-WW1 tanks to mark the spot where the French first used tanks. Still, it offers great information boards. Think of it this way: modern tanks are better than no tanks at all!
   N 49 24.782 E 3 53.266 (49.413033, 3.887767)

2. **VIEUX CRAONNE:** The town of Craonne was destroyed during the battle. It is an important stop from which to appreciate the battle and to see the craters and remains of the town.
   N 49 26.728 E 3 47.385 (49.445467, 3.78975)

3. **PLATEAU CALIFORNIA:** This is where German positions overlooked the fields the French crossed. It's an excellent place to see terrain, trenches and craters.
   N 49 26.694 E 3 46.772 (49.444892, 3.779539)

4. **LA CAVERNE DU DRAGON, "DRAGONS LAIR," CHEMIN DES DAMES:** This is an underground museum at the base of the ridge where the French counterattacked the Germans. You can see the museum only by tour, so plan your visit in advance. Well worth it and very unique.
   N 49 26.510 E 3 43.943 (49.441833, 3.732383)

5. **CERNY-EN-LAONNOIS MEMORIAL CHAPEL AND FRENCH AND GERMAN CEMETERIES:** A chapel is located here at the main crossroads. Across the road are cemeteries for the German and French fallen.
   N 49 26.578 E 3 39.999 (49.442967, 3.66665)

6. **FORT DE LA MALMAISON, GERMAN CEMETERY:** This contains the graves of more than 11,000 Germans who died during WW2. It was a French fort decommissioned prior to WW1. The Germans used it as a strongpoint in the defensive

line during the Chemin Dames offensives. The French captured it after a battle that lasted from October 23-27, 1917, and captured more than 11,000 Germans. That led to the capture of Chemin Dames Ridge, but at a cost of 10,000 French casualties.
N 49 27.565 E 3 31.415 (49.459417, 3.523583)

7. **BASQUE MONUMENT:** This is a monument to the men of the 36th Division, most from the Pyrenees, in honor of their sacrifice in the Battle of Plateau California.
N 49 25.805 E 3 45.583 (49.430083, 3.759717)

8. **FRENCH NATIONAL CEMETERY, CRAONELLE:** Laid out on the slope the French fought their way up during the battle, this cemetery contains the remains of 3,936 combatants, including 24 British graves.
N 49 25.885 E 3 46.391 (49.431417, 3.773183)

BATTLE OF VIMY RIDGE:

9. **VIMY RIDGE MEMORIAL, HILL 142:** If not the best memorial, this is certainly one of the top-5 WW1 memorials/parks to see in Europe. On the ridge dominated by Hill 142, sits a beautiful memorial to the 60,000 Canadians who died in the war. On the base of the memorial are inscribed the names of 11,285 soldiers who have no known grave. On arriving, you can see how the ridge provides an incredible view of the Douai Plain and why they fought over it. The monument parking lot is located at the coordinate:
N 50 22.765 E 2 46.417 (50.379417, 2.773617)

10. **CANADIAN WELCOME CENTER, MEMORIAL, AND PRESERVED TRENCHES:** The Canadian Vimy Ridge Foundation manages the memorial and park, which includes a new information center that provides an excellent overview and artifacts of the battle. The area also includes craters form the battle and the trenches preserved in concrete after the war. The center provides tours of the Grange Tunnel. The parking lot is located at the coordinate.
N 50 22.340 E 2 46.249 (50.372333, 2.770817)

11. **VIMY RIDGE CEMETERIES, CANADIAN CEMETERY NO. 2, AND GIVENCHY ROAD CEMETERY:** These 2 cemeteries are located on the northern edge of the battle, still within the Vimy Memorial Park. Walking among the headstones and reading the quotes provided by families is a moving experience, like so many of the memorials and cemeteries. Canadian Cemetery No. 2 contains the graves of 2,965 dead, and Givenchy Road Cemetery contains 111 graves. From both, you can see the start line and the French memorial on the ridge, Notre Dame Lorette to the north.
N 50 22.615 E 2 45.892 (50.376911, 2.764865)

BATTLES OF ARTOIS / ARRAS:

As described earlier, the Battle of Artois took place over a period of a year. The battles were largely attempts by the French to regain the high ground, such as Notre Dame Lorette, St. Nazaire, and Vimyv Ridge. The places to see are listed under that battle but also included in the Vimy Ridge Map.

NIVELLE OFFENSIVE – 2ND BATTLE OF AISNE, APRIL 16 TO MAY 16, 1917

MAP 14

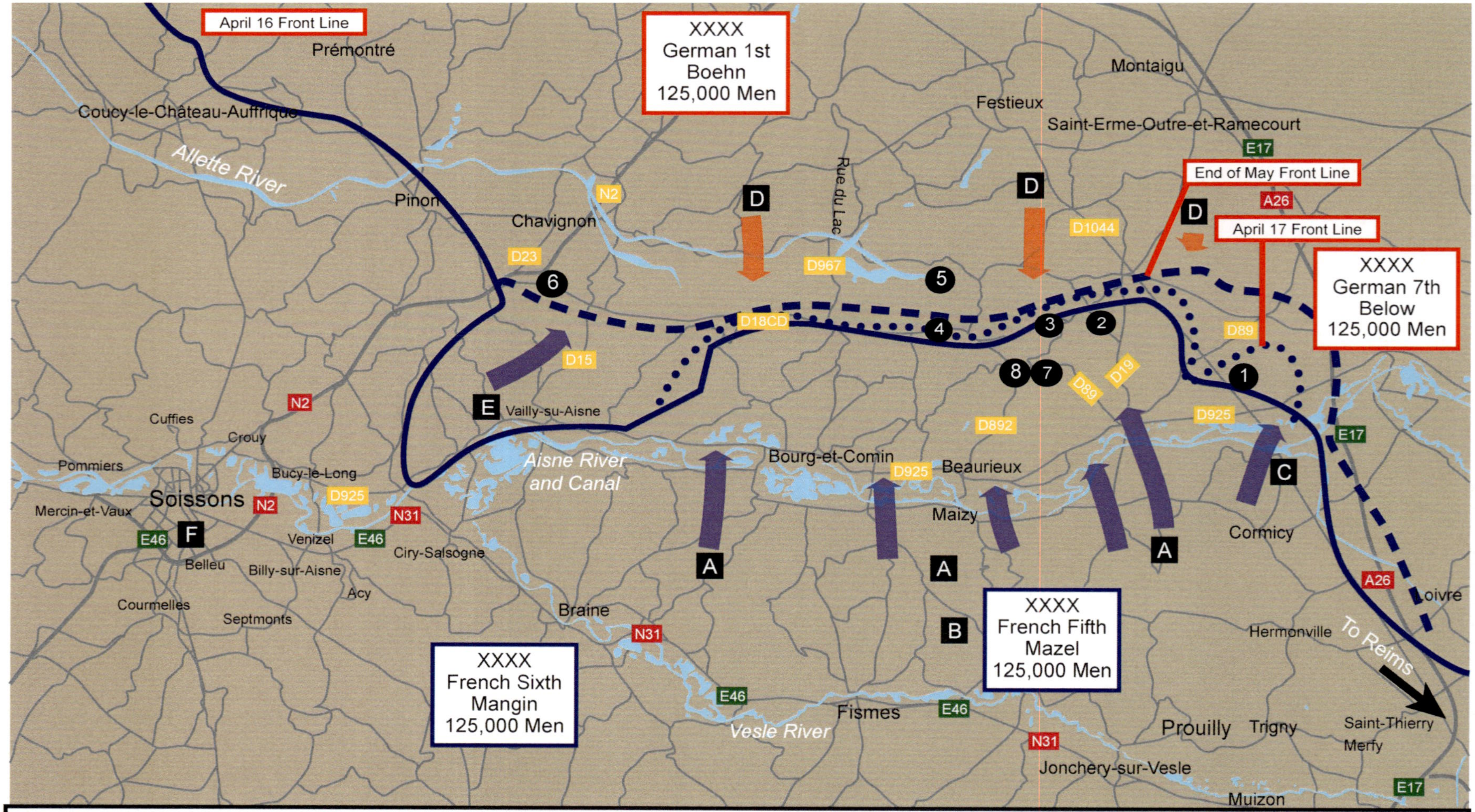

MOVEMENTS

A. On April 16 at 0600 19 Divisions attacked the German positions, aiming to gain 6 miles beyond the ridge.
B. As they attacked, over 3,000 supporting artillery pieces maintained a "creeping barrage".
C. On the right flank, a large force of 200 tanks engaged in the first armor battle for the French.
D. At 1430, the Germans counterattacked the French, because they had prior knowledge about the French Offensive. At the end of the day the French suffered over 40,000 casualties.
E. The first week of May, instead of cancelling the offensive, Nivelle pushed further, with little gain.
F. April 29, the French mutinies began, in response to the carnage.
G. May 15, Nivelle was replaced by Petain and the offensive was stopped.

NIVELLE OFFENSIVE – 2ND BATTLE OF AISNE, APRIL 16 TO MAY 16, 1917

MAP 14

PLACES TO SEE

1. National Monument to Tanks
2. Vieux Craonne
3. Plateau California
4. La Caverne du Dragon Museum
5. Cerny en Laonnois Memorial Chapel and cemeteries
6. Fort de la Malmaison
7. The Basque Monument
8. French Cemetery, Craonne

Photo left is of a German bunker on the Plateau de Californie. Photo below left is looking from the German position to the French lines, towards Reims.

Below right, a photo of the Plateau today looking from almost the same vantage point.

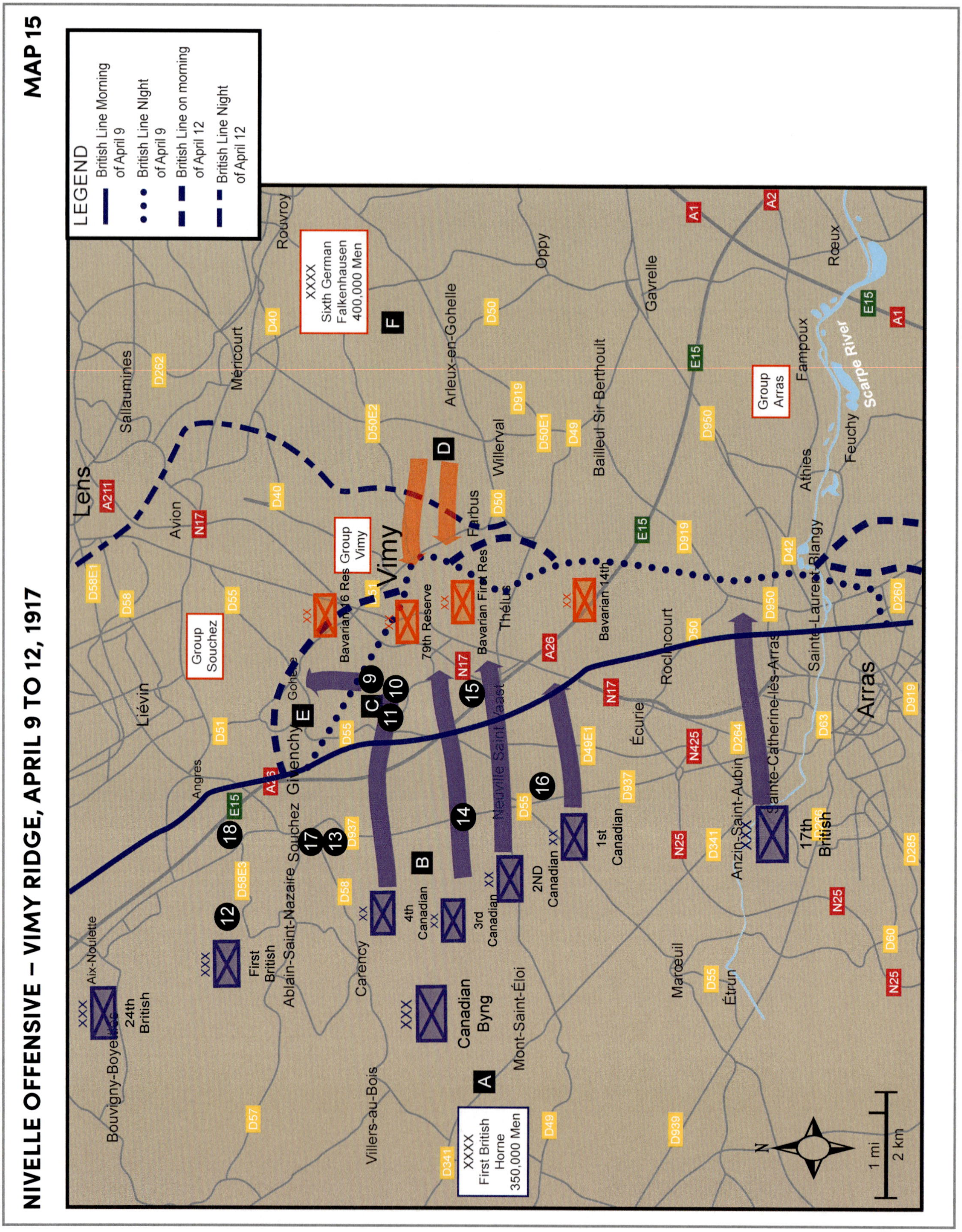
NIVELLE OFFENSIVE – VIMY RIDGE, APRIL 9 TO 12, 1917
MAP 15
LEGEND
British Line Morning of April 9
British Line Night of April 9
British Line on morning of April 12
British Line Night of April 12
XXXX Sixth German Falkenhausen 400,000 Men
XXXX First British Horne 350,000 Men
Group Souchez
Group Vimy
Group Arras
XXX 24th British
XXX First British
XXX Canadian Byng
XX 4th Canadian
XX 3rd Canadian
XX 2ND Canadian
XX 1st Canadian
XXX 17th British
Bavarian 16 Res
79th Reserve
Bavarian First Res
Bavarian 14th
Lens
Liévin
Vimy
Arras
Sallaumines
Méricourt
Avion
Rouvroy
Arleux-en-Gohelle
Oppy
Gavrelle
Willerval
Farbus
Thélus
Bailleul Sir Berthoult
Roclincourt
Écurie
Fampoux
Rœux
Athies
Feuchy
Sainte-Laurent-Blangy
Sainte-Catherine-lès-Arras
Anzin-Saint-Aubin
Marœuil
Étrun
Neuville Saint Vaast
Mont-Saint-Éloi
Givenchy
Souchez
Ablain-Saint-Nazaire
Carency
Villers-au-Bois
Aix-Noulette
Bouvigny-Boyeffles
Angres
Scarpe River
1 mi
2 km
N

NIVELLE OFFENSIVE – VIMY RIDGE, APRIL 9 TO 12, 1917

MAP 15

MOVEMENTS

A. Starting March 20, the Allies shelled the German's on a 24 mile front, in total firing over 2,680,000 rounds, including gas shells for the last 10 hours.

B. April 9 at 0530, in the driving sleet the Canadians attacked from their trenches and tunnels under the protection of a creeping barrage.

C. At 1300 the Canadians captured Vimy Ridge.

D. For 2 days, the Germans attempted several counterattacks but the Canadians beat them off.

E. April 12, the Canadians took the town of Givenchy in the north. The attack worked as planned for a deception of the Nivelle Offensive in Champagne, the Germans moved 2 times the amount of troops to defend from a bigger break through to the Vimy area, leaving Champagne lightly defended.

PLACES TO SEE

9 Vimy Ridge Memorial, Hill 142
10 Vimy Ridge Memorial Center and Trenches
11 Vimy Ridge Cemeteries
12 Notre Dame Lorette Memorial, Museum and Ring of Rememberance
13 Cabaret Rouge Cemetery
14 Memorials to Polish and Czechoslovakia Volunteers
15 LIchfeld Crater Memorial
16 German Cemetery Arras
17 General Barbot, 77th ID Monument
18 Lens Center for Peace, 14-18

To the left, Canadian artillery firing during the Battle of Vimy Ridge. Photo right is a German MG in action. (NARA)

Photo left, is the statue of "Canada" mourning over her fallen at the Vimy Memorial, overlooking the fields beyond to the east. Photo right is the entrance to the memorial and the cratered landscape, the memorial is on Hill 142.

Further along the ridge, the Canadians have a new welcome center, with the above preserved trenches (preserved in the 1930s for the memorial). The photo right is of the small Givenchy Road Cemetery from the optic of the German line. It is one of two cemeteries on the ridge.

Photo left is of the Canadians digging in on Vimy Ridge after the battle. Their success was due to excellent planning which included plans to dig in and repulse the guranteed German counterattack. (NARA)

Just NW of Reims is the French National Tank monument to honor the location where the French tanks were first used in battle. The upper left photo, Schneider tank was used but to little effect.(NARA) In all, they committed 128 tanks and 118 were lost. The middle photo is of one of the armored vehicles at the memorial (1950s era). Photo right, one of the many crucifixes in the countryside.

Photo left, the town of Craonne on the ridge, what is left of it, after the battle. Center and right photos are on the ridge, in the center called "Plateau California, excellent stop to view the center of the battle and memorials, to include the artist memorial, photo right.

Left photo, entrance to the excellent underground museum, Dragon's Lair. Center photo is the Cerny en Laonnois Chapel with the French Cemetery located across the street from it.

To the left, a photo of the Chemin Dames ridge being shelled duringthe battle, from the perspective of the French. One can see that the dominate high ground would prove to be a very difficult obstacle, making an attack costly. (NARA)

BATTLES OF BULLECOURT

April 11, 1917, May 3-17, 1917

"I hope the war will soon be over because it is destroying the best men and everything that is beautiful and civilized in life."
—Australian Lt. Wilfred Barlow, KIA May 12, 1917

SUMMARY: At Bullecourt in April and May, 1917, the Allies again threw good men into a death trap for no significant goals or gains. Earlier, in March, the Germans had withdrawn to the Hindenburg Line, and the Allies were slow to react and had some skirmishes with the German rear guard. In early April, the Nivelle Offensive began with the diversionary Canadian-led attack on Vimy Ridge, followed by the larger attack in Chemin Dames, northwest of Reims, Bullecourt is located between the two.

Some could conclude that the German defenses around Bullecourt were weakened while the Battles of Vimy and Battle of Chemin Dames were commencing. The goal was for Australian and British units to break through and capture part of the Hindenburg Line, one of the first "tests" of the line. The planning was so poor the British 5th Army commander, General Hubert Gough, did not bother with an artillery preparatory barrage. Instead, on April 11 at 0445 after a night of huddling to stay warm during a blizzard, the 4th and 12th Brigades of the Australian 4th ID advanced across open fields to capture the part of the Hindenburg line lying between the towns of Bullecourt and Queant. The 4th Brigade was on the right, the 12th on the left. The 12th waited for the 11 tanks that were supposed to support their attack and help break through the barbed wire, but most did not reach the staging point because of mechanical issues. Meanwhile, the 4th Brigade moved forward. Despite the 30-meter-deep barbed wire, artillery, rifles, and machine guns, the brigade penetrated the German 1st line.

Then the Australians pushed 2 more battalions to pass through the 4th's position and attacked the German 2nd line, but they were cut down and stopped cold. After about 6 hours of fighting, the Germans counterattacked at 1000. Units of the German 27th Wurttemburg Division came out of their shelters well-rested and quickly overran the remaining Australians. It was devastating. The Australian 4th Brigade of about 3,000 men suffered 2,229 casualties. The 12th lost a little over 1,200. The Germans suffered more than 750 KIA.

Smelling opportunity, the Germans counterattacked on April 15 and pushed the Australian line back about 1 to 2 km. Over the next week or so, the Allies attacked and recaptured that terrain, and on May 3, the Nivelle Offensive in Chemin Dames was failing. So, the commanders ordered another attack at Bullecourt, this time with better planning and artillery, yet still a hard target.

On May 3 at 0345, after a preparatory barrage, 4 waves of infantry attacked the Hindenburg Line, following a rolling barrage. The direction of attack was similar and so were the results. The Australian 2nd Division led the attack, with the 5th Brigade on the right and the 6th on the left. The 5th was decimated and retreated, but the 6th was able to advance 400 meters and capture the 1st line. That night, they received reinforcements and began digging a communication trench back to the main line to move wounded and ammunition. After another attempt to advance, however, they had to withdraw.

On the left flank, the British 62nd West Riding ID attacked Bullecourt with little gain and high losses. At the end of the first day, the 62nd was replaced by the British 7th ID, and the Australian 2nd ID was replaced by the Australian 1st ID. Next, the British attacked on May 7 with the aid of the Australian 5th ID and they were able to capture most of the village by May 12. The Germans counterattacked and hard fighting continued for the village until May 15 when the Germans began to withdraw. On May 17, Bullecourt was finally captured and in Allied possession. But, it was only a bump into the first section of the Hindenburg Line. It did not provide any strategic value and didn't threaten the line, because another attack would have been on a narrow front surrounded on three sides. The May battle caused additional staggering losses. The Australians had more than 7,000 casualties, the British more than 2,700. The Germans suffered more 6,000 casualties.

OUTCOME: No further gains, only major losses in life and limb.

WHO? The British 5th Army was made up of British and Australian soldiers. Of the units from the 7 divisions committed, the Australians suffered more than 10,000 casualties and the British close to 4,000. With the aid of the Hindenburg Line defenses, the Germans only needed 1 division with supporting reserves to defend the sector, suffering about 7,000 casualties.

WHEN? April 11 and May 3-17, 1917

WHY AT BULLECOURT? It was believed that with the attacks at Vimy Ridge and in Chemin Dames during the same period, there was a chance to break through the "new" Hindenburg Line at this location.

PLACES TO SEE:

1. **BULLECOURT CHURCHYARD MEMORIAL:** At this location are memorials to the divisions involved in the battle. From here you can walk about 50 yards to the northeast to the farm road, "Rue des Australians." It is a decent short hike worth the walk to memorials. This road followed the German defenses.
   N 50 11.578 E 2 55.787 (50.192975, 2.929785)

2. **AUSTRALIAN MEMORIAL PARK:** Here is a statue and memorial to the soldiers who fought and died here. The park is located on the 1st row of the German Hindenburg Line.
   N 50 11.685 E 2 56.291 (50.194745, 2.938177)

3. **MEMORIAL CROSS:** Farther ahead is a small cross dedicated to those in the battle with no known graves. It also marks the German 2nd line of defense.
N 50 11.719 E 2 56.629 (50.195312, 2.943820)

4. **GERMAN POSITION, "SIX CROSS ROADS":** At this location is a crossroads that provide a good perspective from the German position during the battle. When observing, keep in mind the machine gunners would typically reach a range of about 800 meters.
N 50 11.810 E 2 56.986 (50.196831, 2.949768)

5. **SUNKEN ROAD, AUSTRALIAN POSITION:** This spot was mid-point of the attack route by the Australian 12th Brigade. The soldiers used the "sunken road" for some protection from the withering fire. Standing here, you can appreciate the distance the soldiers had to traverse under fire and with no cover.
N 50 11.366 E 2 56.201 (50.189436, 2.936690)

6. **BULLECOURT 1917, MUSEE JEAN ET DENISE LETAILLE:** An excellent Bullecourt museum in the small village, with an abundance of local battlefield discoveries. As of 2018, the museum is only open on weekdays.
N 50 11.513 E 2 55.633 (50.191889, 2.927224)

BATTLES OF BULLECOURT, APRIL 11 AND MAY 3 TO 17, 1917

MAP 16

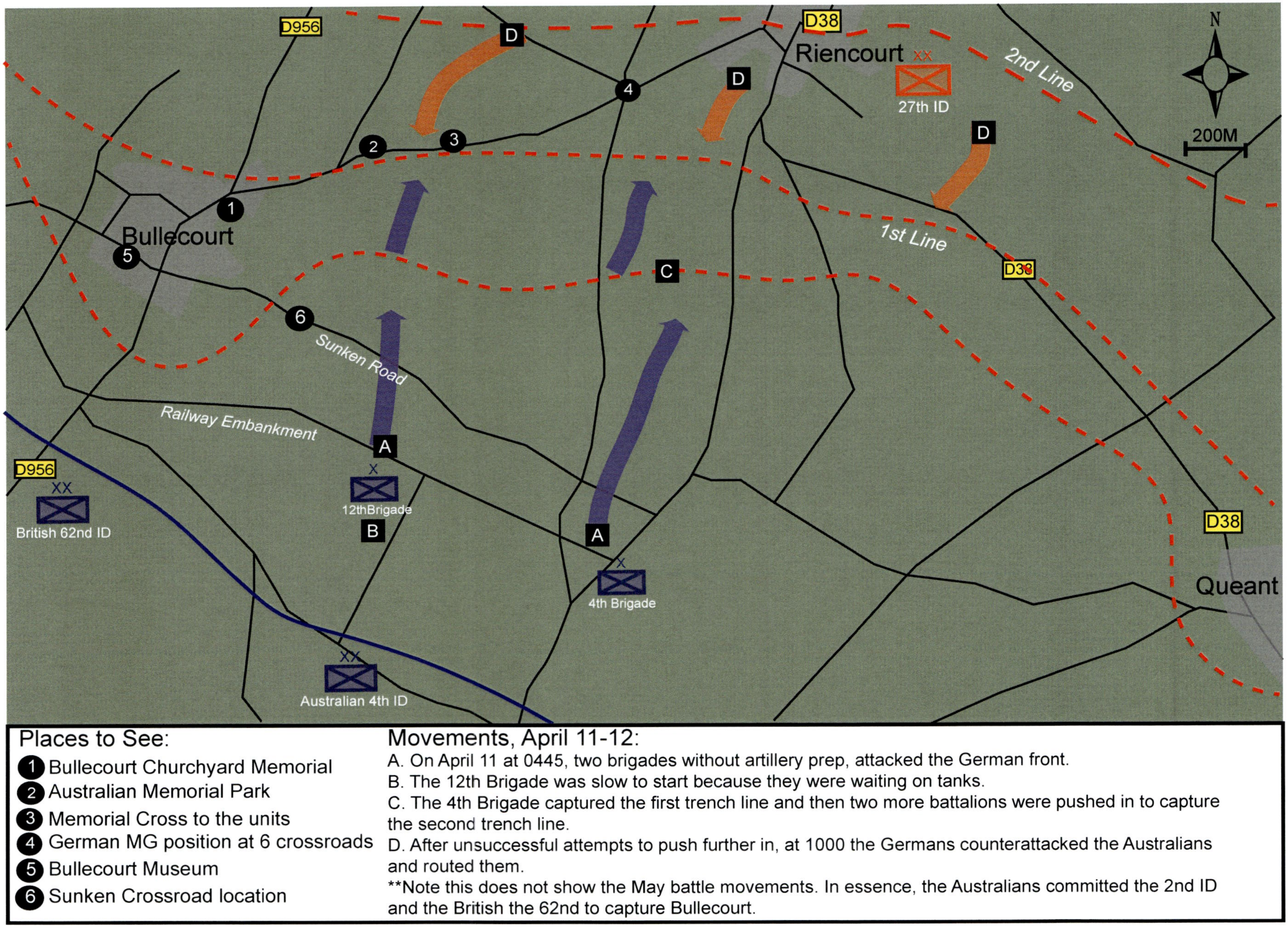

Places to See:

1. Bullecourt Churchyard Memorial
2. Australian Memorial Park
3. Memorial Cross to the units
4. German MG position at 6 crossroads
5. Bullecourt Museum
6. Sunken Crossroad location

Movements, April 11-12:

A. On April 11 at 0445, two brigades without artillery prep, attacked the German front.
B. The 12th Brigade was slow to start because they were waiting on tanks.
C. The 4th Brigade captured the first trench line and then two more battalions were pushed in to capture the second trench line.
D. After unsuccessful attempts to push further in, at 1000 the Germans counterattacked the Australians and routed them.
**Note this does not show the May battle movements. In essence, the Australians committed the 2nd ID and the British the 62nd to capture Bullecourt.

BATTLE OF CAMBRAI

November 20-December 7, 1917

"The Infantry felt that he could do practically nothing against its armored sides."–Field Marshal Paul von Hindenburg, November, 1917 Journal

---

SUMMARY: The Battle of Cambrai was a major British offensive in an attempt to break through the formidable German Hindenburg Line near Cambrai, France. If successful, the British would threaten the German northern flank and supply routes. The battle is known for its large commitment of the British Mark IV tanks (more than 470), which were somewhat effective. Historically, the tank commitment overshadows the very extensive planning and successful application of a combined arms offensive with infantry, tanks, artillery, and aircraft. The battle gained some ground for the British, but it proved costly when the Germans counterattacked and recaptured the majority of terrain, using stormtrooper tactics.

The Hindenburg Line was a formidable section of the German front line, extending kilometers in depth with concrete strongpoints, laced with barbed wire and trenches. Plus the defensive line incorporated two canals, the St. Quentin and Canal du Nord, running north to south, and towns which created more obstacles. In addition, the Germans benefited from there having been no major attacks in the sector for almost 3 years. That allowed them time to fortify and improve the defenses. All the same, the British still wanted to attack the sector, believing they could threaten the German supply line and cut off its northern armies.

In great secrecy, the British decided on a combined arms attack with artillery, aircraft, infantry, and tanks. For artillery, they committed more than 1,000 guns and used sound-ranging to register their aim versus firing test rounds that would have given away their locations and an obvious sign of imminent attack. Also, they decided on infantry infiltration tactics that would follow 300 yards behind a rolling barrage from the artillery, with tanks in support to drive over barbed-wire obstacles. The Royal Flying Corps also deployed for support after new training in a ground-attack role to attack trenches, supply points, and airfields. The British committed 2 corps comprising 8 divisions (2 in reserve) and 470 tanks. On the left flank, IV Corps containing the 36th and 62nd IDs, and 51st Highlanders was in place with tank support. III Corps on the right flank/south contained the 6th, 20th, and 12th IDs. The British initially faced 1 German corps containing the 54th ID, 9th Reserve Division, 183rd Division and 79th Reserve Division.

On November 20 at 0630, more than 1,000 cannons bombarded the German lines and fired smoke rounds, and soon after fired a rolling barrage ahead of the infantry. Even though the British attempted to keep the attack secret, the Germans were ready for "an" attack. The combined-arms tactics still worked, however, and most of the British attack gained 8 km. In the south, the 20th ID captured La Vacquerie, the 12th reached Lataeu Wood, and the 6th captured Ribecourt and Marcoing. In the north on the left flank, the 51st Highlanders were held up after losing 40 tanks trying to capture Flesquieres. The 62nd ID on the left of the 51st battled through Havrincourt and Graincourt. The 36th Ulster Division attacked down the Bapaume Road to Cambrai Road rather successfully. By the end of the day, the British had suffered 4,000 casualties (which was low), and lost 65 tanks to the enemy, 71 others for mechanical reasons and 43 that were ditched because they were stuck or damaged. And, they captured more than 4,000 Germans. Overall, the British gained significant terrain, but did not reach the northern flank objective of Bourlon Ridge.

The next day, German reinforcements began arriving and planning a counterattack. On the 2nd day, the British halted the right flank's III Corps operations and focused on taking Bourlon Ridge. After fending off a German counterattack, the 62nd ID was battle-worn and replaced on November 23 by the 40th ID. That same day, the 40th, with the support of 100 tanks and 430 artillery pieces, attacked the ridge, but only reached the edge of the woods. After 2 more days and 4,000 casualties, it reached the ridge crest.

On November 27, the 62nd ID returned with 30 tanks to replace the 40th. The Germans counterattacked again, but were repulsed after fierce fighting. The British position was precarious and their position created a salient jutting out of the line that measured 11 by 10 km. On November 28, the British dug in on the ridge and were hammered with 16,000 artillery rounds.

On November 30, the Germans had received 20 divisions as reinforcements. At 0700 after a short barrage, the Germans attacked across the front with 9 of those divisions. Using stormtrooper tactics with heavily armed soldiers who bypassed strongpoints, they quickly pushed forward 13 km. On Bourlon Ridge, the British repulsed the German attacks, pouring deadly machine-gun fire on the exposed infantry. Also, the British committed tanks and an infantry division to counterattack the German assault. It was not enough, and on December 3, the British had to withdraw back over the St. Quentin Canal. On December 7, they were forced to withdraw farther, but even though having lost Bourlon Ridge, they still held the captured towns of Havrincourt, Ribecourt, and Flesquieres.

OUTCOME: The British gained some ground, retaining 3 towns on the northern/left flank, while the Germans gained ground on the southern flank. The Hindenburg Line remained intact and the German overall success fueled confidence that they could break through in the spring before the Americans arrived in force.

WHO? The British committed a combined-arms approach with 2 Corps and 476 tanks under the command of General Haig. They suffered 44,000 casualties and lost approximately 175 tanks. The Germans, under the shelter of the Hindenburg Line, committed 1 corps of troops initially and suffered 45,000 casualties.

WHEN? November 20, 1917-December 7, 2017.

WHY IN CAMBRAI? The British attacked this largely quiet sector of the line, which had not seen much fighting since 1914. It was an attempt to cut off the German northern armies and threaten their supply chain.

PLACES TO SEE:

1. **CAMBRAI WAR MEMORIAL, LOUVERAL, AND CEMETERY:** This British memorial is on the far left flank of the British attack. It commemorates the 7,058 British and South African soldiers and airmen who died during the November-December battle of Cambrai. The British Commonwealth cemetery contains 124 graves from the battle.
   N 50 08.201 E 3 00.929 (50.136682, 3.015481)

2. **HAVRINCOURT, 62ND WEST RIDING ID MEMORIAL:** Havrincourt was one of the initial objectives of the British and on the front line. The 62nd ID captured and liberated Havrincourt and participated in the attack on Bourlon Ridge.
   N 50 06.520 E 3 05.286 (50.108660, 3.088096)

3. **FLESQUIERES MEMORIAL AND INFORMATION BOARD:** This is a good place to start and get your bearings. The town of Flesquieres was well-defended by German artillery acting as anti-tank guns. More than 40 British tanks were destroyed trying to attack the town on the first day. That night, as a tactical retreat, the Germans withdrew, and it was occupied by the British the next day. Today, the town created a memorial with a concrete pad that has the marks of tank tread and soldiers feet. The information board is excellent and the location provides a great panorama of the battlefield. There is also a hiking path across the road that leads through the farm land to Havrincourt.
   N 50 07.194 E 3 07.330 (50.119901, 3.122166)

4. **FLESQUIERES HILL CEMETERY AND MK IV TANK "DEBORAH":** Historian and author Philippe Gorcynski researched for 6 years to find a tank buried during the Battle of Cambrai. In 1998, after thorough research and scanners, under 15 feet of soil he located a buried MK IV tank designated D51 and named "Deborah." It was an MK IV tank "female," which had 6 machine guns, unlike the "male," which had 4 machine guns and a 57mm cannon. The tank led an attack on Flesquieres through the village, and was knocked out at the edge of town by direct fire from German cannons. The action killed 4 of the crew. The tank commander, 2nd Lt. Frank Heap, was awarded the Military Cross. The citation reads: "In Cambrai operations near Flesquieres on November 20th 1917, he fought his tank with great gallantry and skill, leading the infantry on to five objectives. He proceeded through the village and engaged a battery of enemy field guns from which his tank received five direct hits, killing four of his crew. Although then behind the German lines he collected the remainder of his crew and conducted them in good order back to our own lines in spite of heavy machine gun and snipers fire". In March, 1918, the Germans took over the Flesquieres Hill again and pushed the tank into a crater to use it as a shelter. As of March, 2018, the tank was moved to a new location next to Flesquieres Commonwealth Hill Cemetery, just east of town in a beautiful museum. The cemetery contains more than 900 graves

of Commonwealth soldiers, including the D51 tank crew.
N 50 07.382 E 3 07.138 (50.123037, 3.118967)

5. **BOURLON RIDGE TANK MEMORIAL:** Bourlon Ridge is located on the northern edge of the battlefield, and is critical high ground that dominates the surrounding fields. The British forces fiercely fought for the terrain and only after thousands of casualties did they dislodge the Germans and hold the ridge. After holding the ridge for only about 10 days, they withdrew because of the German counterattacks. At this location is an information board and small monument to the men and tanks that fought during the battle. The British initially used 20 tanks to attack the Germans, but because of battle damage and mechanical issues, only 5 made it into the fight. You can also park and hike around the ridge, or wait until you drive up a little farther to the top where the Canadian memorial is.
N 50 10.427 E 3 06.959 (50.173789, 3.115975)

6. **BOURLON RIDGE, CANADIAN MEMORIAL, TRAIL, AND COMMONWEALTH CEMETERY:** Here is an impressive Canadian memorial overlooking the town and fields below. It is a memorial to the Canadians who fought and died fighting for Bourlon Ridge during the 100 Days Offensive that assisted in ending the war, from September-November, 1918. The Canadian efforts allowed other divisions to capture Cambrai and surrounding towns. In all, they lost 20,000 men in the last days of the war. I recommend also walking along the public path to explore the woods, fox holes, and craters. To the north is the Bourlon Woods Commonwealth Cemetery, which contains nearly 250 graves.
N 50 10.445 E 3 07.097 (50.174091, 3.118284)

7. **FONTAINE NOTRE DAME INFORMATION BOARD:** The Cambrai Battlefield has approximately 20 excellent information boards. I highlight this location because it was on the front line at the end of the battle, and another significant location. Most of these are located at the center of the main towns and near the WW1 French town memorials.
N 50 09.986 E 3 09.551 (50.166433, 3.159178)

8. **ANNEUX, COMMONWEALTH CEMETERY:** Anneux was an objective captured in the 1st days of the battle by the 62nd ID. It was also a major battleground during the German counterattack, finally recaptured by the Germans in early December. In the area are many British cemeteries. This one I highlight because of the 1,013 graves with more than 450 unidentified, an indicator of the severity of the battles.
N 50 09.583 E 3 07.602 (50.159718, 3.126699)

9. **FORT GARRY HORSE MEMORIAL, MASNIERES:** Along the canal to the east of the large French Memorial and Battle of Cambrai Information board is a stone memorial to the Canadian Cavalry Unit, Fort Garry Horse. During the Battle of Cambrai, they were assigned to rush through the gaps created by tanks and capture German HQ units in the rear. On November 20, 1917, after finding the main bridge damaged by a tank, they crossed the canal at 1530. The cavalry squadrons split and attacked German positions on horseback. They broke through, taking

out machine gun and artillery positions, but eventually had to return because no one was on their flanks. Of the 133 cavalry who attacked, only 46 returned. Lt. Hew Strachan received the Victoria Cross. His citation reads: "On 20 November 1917 at Masnières, France, Lieutenant Strachan took command of a squadron of his regiment when the squadron leader, approaching the German front line at a gallop, was killed. Lieutenant Strachan led the squadron through the enemy line of machine-gun posts and then, with the surviving men, led the charge on the German battery, killing seven of the gunners with his sword. When all the gunners were killed and the battery silenced, he rallied his men and fought his way back at night through the enemy's lines, bringing all unwounded men safely in, together with 15 prisoners."
N 50 06.744 E 3 12.863 (50.112408, 3.214390)

10. **MASNIERES NEWFOUNDLAND MEMORIAL:** This location has the Canadian Caribou Memorial commemorating the actions of the Royal Newfoundland Regiment during the Battle of Cambrai. The regiment was one of four battalions in the British 29th ID. The town of Masnieres was a strongpoint in the Hindenburg Line, and was captured after tank, aircraft, artillery, and infantry attacks.
N 50 07.405 E 3 12.715 (50.123414, 3.211909)

11. **TRESCAULT, 42ND TERRITORIAL DIVISION MEMORIAL:** Here there is a memorial to the British 42nd TD, which fought in September 1918 offensive to recapture the land and fight through the Hindenburg Line.
N 50 05.566 E 3 06.046 (50.092774, 3.100774)

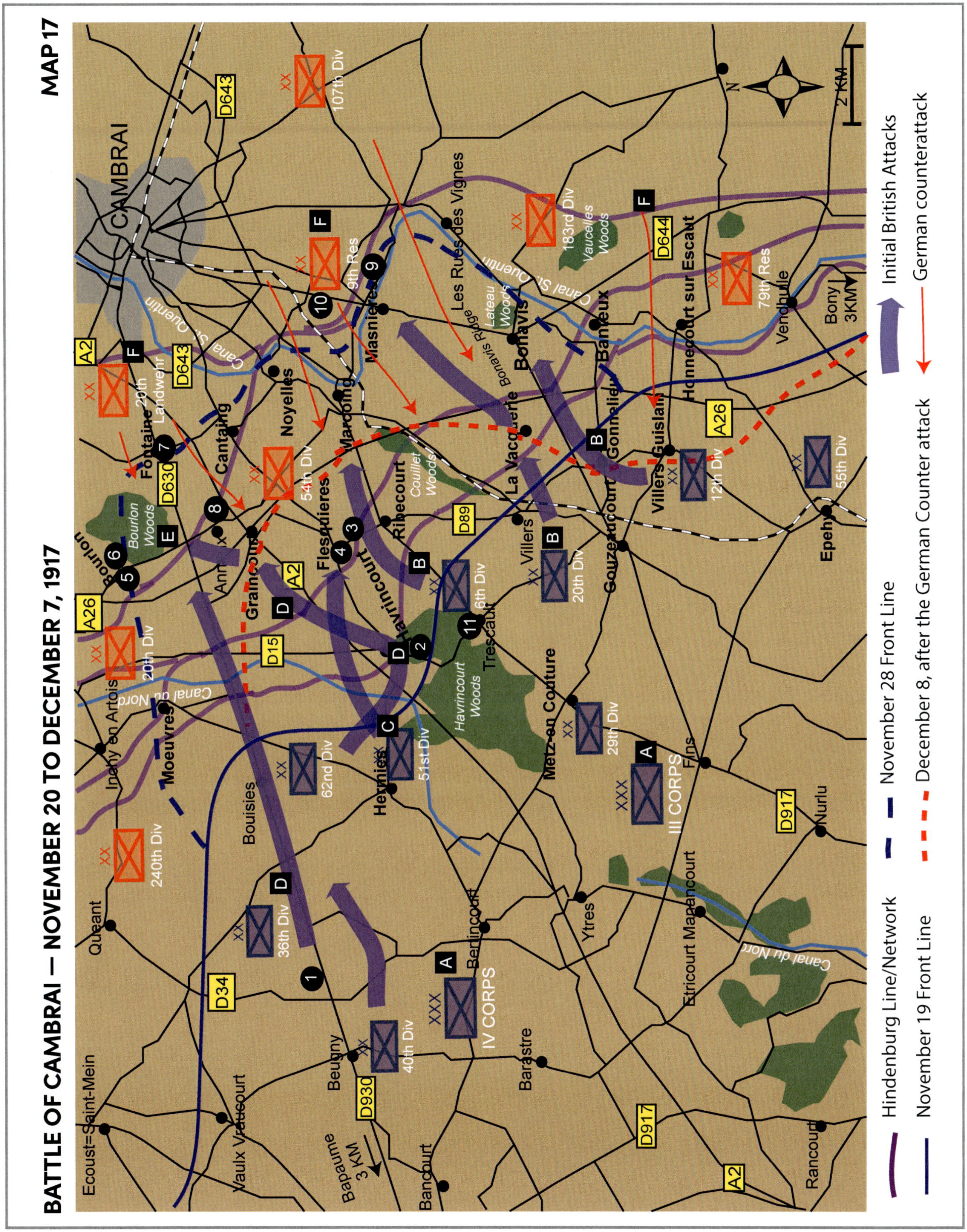
BATTLE OF CAMBRAI — NOVEMBER 20 TO DECEMBER 7, 1917
MAP 17
CAMBRAI
Canal St. Quentin
Canal du Nord
Bourlon Woods
Havrincourt Woods
Couillet Woods
Lateau Woods
Vaucelles Woods
Bonavis Ridge
IV CORPS
III CORPS
107th Div
183rd Div
9th Res
79th Res
20th Landwehr
54th Div
20th Div
240th Div
36th Div
40th Div
62nd Div
51st Div
6th Div
29th Div
12th Div
55th Div
Hindenburg Line/Network
November 19 Front Line
November 28 Front Line
December 8, after the German Counter attack
Initial British Attacks
German counterattack
2 KM
Bapaume 3 KM
Bony 3KM

BATTLE OF CAMBRAI — NOVEMBER 20 TO DECEMBER 7, 1917 — MAP 17

MOVEMENTS:

A. On November 20 at 0630 over 1,000 British cannons opened fire on German lines, first with HE, then smoke rounds and followed by a rolling barrage in front of the infantry.

B. The 20th ID captured Vacquerie, the 12th reached Lateau Wood and the 6th ID captured Ribecourt and Marcoing.

C. In the North, the 51st Highlander Division, after severe tank loses captured Flesqueries on November 21, after the Germans withdrew.

D. The 62nd ID captured Havrincourt and Graincourt and the 36th Ulster Division made gains close to Bourlon Ridge.

E. Over a three day period, the British battled to capture Bourlon Ridge, using the reserve 40th ID, grinding them down and then throwing the 62nd ID back into the fray to capture the ridge on November 23, over the next few days the British threw back several German counterattacks.

F. On November 30, the German 20 reserve divisions had arrived for a counterattack. At 0700 the Germans using storm trooper tactics attacked. The British withdrew to St. Quentin Canal on December 3 and had to withdraw further by December 7.

PLACES TO SEE:

1. Louveral, Cambrai Memorial and Cemetery
2. Havrincourt, 62nd ID West Riding Memorial
3. Flesquieres Battle Memorial and Info Board
4. Flesquieres, Deborah Tank Display
5. Bourlon Ridge, Tank Memorial
6. Bourlon Ridge, Canadian Memorial, Trails and Cemetery
7. Fointaine Notre Dame, Information Board
8. Anneux British Cemetery
9. Fort Garry Horse Memorial, Masnieres
10. Newfoundland Memorial, Masnieres
11. 42nd TD Memorial, Trescault

Above a German photo of a captured British tank and below left, a British tank that suffered several deadly hits.

Photo left, the author with the tank "Deborah" in the new museum. The tank was discovered by Philippe Gorcynski in the late 1990s. As of March 2018, it is housed in a new museum, next to the Flesquieres Cemetery, where the crew is buried. Photo right is of the Flesquieres tank monument at the eastern edge of Flesquieres, which was at the center of the battle in 1917.

Photo lleft, was taken from the information board and map at the Flesquieres memorial, looking west from the perspective of the Germans. Photo right is on the road to the north, in the view of the British looking east, definitely good tank country.

On the northern flank of the British on their attack towards Cambrai was Bourlon Wood. The photo left is from the Canadian memorial on Bourlon Wood, the photo is taken from the German perspective. The British fought for this dominating hill and woods in 1917, then the Canadians fought for it in 1918. Today, you can walk the woods and still see the craters and trenches. Photo right, is a memorial to the tanks that assisted in taking the hill in 1917.

Photo above is of the British Bourlon Cemetery just down the dirt path from the Canadian Memorial. Photo right is a destroyed British tank from the Battle of Cambrai. (BA)

GERMAN FINAL OFFENSIVES

1918

"If we are victorious in one more battle with the Romans, we shall be utterly ruined."—Plutarch quoting Greek King Pyrhhus, and the origin of the phrase, Pyrhhic victory.

SUMMARY: Ironically on November 11, 1917, one year prior to the capitulation of the Germans in 1918, High Commander Erich Ludendorff decided to go on the offensive in the Spring of 1918. Several factors contributed to that decision. In 1917, in Riga against the Russians and in Caporetto against the Italians the Germans were very successful using large stormtrooper tactics, in both battles having major breakthroughs. Also the Allied blockade had been impacting the German food supply and the Germans believed a famine could happen. The Russians had just surrendered in the east to the Germans which freed up 50 divisions (men and materiel) giving the Germans superior numbers in the west for the first time since 1914. Finally, the other major contribution was that Ludendorff wanted to strike a blow before the Americans were fully engaged in the war. Ludendorff understood that the American superior resources in men and materiel would eventually defeat the Germans. Ludendorff's strategic objective was to hit the Allies hard enough and multiple times to split the French and British forces, isolate the British, knock them out of the war and bring the Allies to the negotiating table.

The Germans focused on breaking through the British line in the north, with several attacks in the south to draw reserves and assistance away from the north. In total, the 5 attacks from March through July, 1918 gained ground, but their artillery and logistics could not keep up with the forces. The German soldiers were slowly worn down and armies stuck in salients, surrounded on three sides by the Allies. As a result, the advances were pyrrhic victories. They simply lost too many well-trained and equipped men. The following are brief summaries of each of the main battles:

OPERATION MICHAEL – SOMME OFFENSIVE; MARCH 21-APRIL 5, 1918

On March 21, the Germans attacked between the British and French forces in the Somme to capture the main logistics hubs of Amiens and Arras. At 0440, the Germans fired more than 1.1 million artillery rounds in a 5-hour period, covering a span of 150 square miles, one of the largest battle areas of the war. The morning was foggy, aiding the German stormtrooper tactics in unison with the rolling artillery barrage, called "feuerwaize" (rolling fire). The Germans attacked with 3 armies and an additional corps against the British 5th Army. The British reeled back, suffering more than 7,500 KIA and 10,000 wounded on the first day, but they and the Australians held their positions around their supply hubs of Amiens and Arras. On March 28, Ludendorff ordered Operation Mars, an attack on Arras which was repulsed by the British. Meanwhile, the German artillery and supplies could not keep up with their advancing soldiers and the Allies sent in reinforcements on the flanks of the German attack. On April 5, the Germans called off the attack. The Germans gained large swaths of land, but it was battle-scarred for them and hard to defend. Both sides suffered tremendously, the Allies more than 255,000 casualties, 1,300 artillery pieces, and 200 tanks lost. The Germans suffered just over 239,000 casualties, approximately 16,000 per day of battle.

OPERATION GEORGETTE – BATTLE OF LYS; APRIL 9-APRIL 29, 1918

Operation Michael had stressed the British reserves and weakened the defenses guarding the channel ports. The Germans decided to attack the British logistics hub of Hazelbrouck and targeted the ports to isolate and starve the British of supplies. The Germans chose their route wisely, and on April 9 attacked the sector defended by the battle-worn Portuguese. The Portuguese had 2 divisions that were about to be replaced by the British. One of the divisions had already left, leaving the other to defend an entire 10-km front until the British replacements were in position. The Germans attacked with 8 divisions and quickly destroyed the Portuguese division, which although resisted bravely, was annihilated. On the Portuguese northern flank, the British 40th ID also collapsed, but on the southern flank, the British 55th ID held. By mid-April, the Germans were fewer than 25 km from the ports and reaching their objectives. Yet again, the logistics and artillery could not keep up with the advance and Allied reinforcements stymied further advance. The Germans called off the attack on April 29 after suffering more than 110,000 casualties. The Allies suffered about the same number.

OPERATION BLUCHER – YORCK – THIRD BATTLE OF AISNE; MAY 27-JUNE 6, 1918

The northern operations were believed close to success, although the Germans needed to draw Allied reserves away from there. Therefore, the Germans conducted this operation that attacked French forces on the battle-worn clay fields between Soissons and Reims. Unfortunately, 6 British Divisions were in the line, resting after being depleted during Operation Michael. Their defensive line did not have depth and was rife for failure. The Germans opened up with a 4,000 cannon rolling barrage, breaking through the defenses and, to their surprise, they reached the Marne River with the possibility of reaching Paris. Within three days they netted 50,000 Allied soldiers. Yet again, the artillery and logistics could not keep up with the German advance, and the Allied reinforcements arrived to blunt it. The attack was called off on June 6 after the Germans suffered 130,000 casualties and the Allies more than 137,000.

OPERATION GNEISENAU – NOYON OFFENSIVE; JUNE 8-12, 1918

Because of the gains in the north, the Germans had two salients that were exposed to artillery fire on three sides and required additional divisions to defend the extended front line. Operation Gneisenau was an effort to straighten the line and remove the gaps. The German attack began on June 8 along the Matz River towards Compiegne. Twenty-one German divisions advanced 9 miles into the Allied lines. Then on June 11, the French counterattacked the Germans with 4 divisions and 150 tanks at Compiegne, stopping their advance. The Germans called off the attack after suffering 30,000 casualties to the Allies' 35,000.

OPERATION MARNESCHUTZ – REIMS / FRIEDENSTURM – 2ND BATTLE OF THE MARNE; JULY 15-18, 1918

Even after the severe losses of irreplaceable stormtroopers, the Germans saw a chance of success. They wanted to attempt a crossing of the Marne River, threaten Paris, and force the French to move more reserves to the area, leaving the British weakened in the north. After that "success," the Germans would attack in the north with Operation Hagan to knock out the British.

On July 15, the Germans attacked with 3 armies to the west and east of Reims. The French held their ground to the east. West of Reims, the French faltered and the Germans crossed the Marne, nearing a breakthrough on the road to Paris. On July 18, however, a preplanned Allied counterattack was launched to threaten the German salient. The counterattack forced the Germans back and to withdraw their forces to more-defensible positions. During the devastating battle, the French suffered 95,000 casualties, the British 17,000, Americans 12,000, and Italians 9,000. The Germans suffered a total of 140,000 casualties and 30,000 captured.

OUTCOME: Not good for the Germans. The German economy and people were suffering from an embargo and could not replace their losses from the failed offensives. Once the Americans joined the battle and reinvigorated the Allies, the Germans would never be able to fill the gaps in the army to be able to win. After July 18, the Allies were took the initiative and the Germans struggled to maintain military and political cohesiveness, resulting in capitulation in 4 months.

SUMMARY GERMAN SPRING OFFENSIVES, 1918

MAP 18

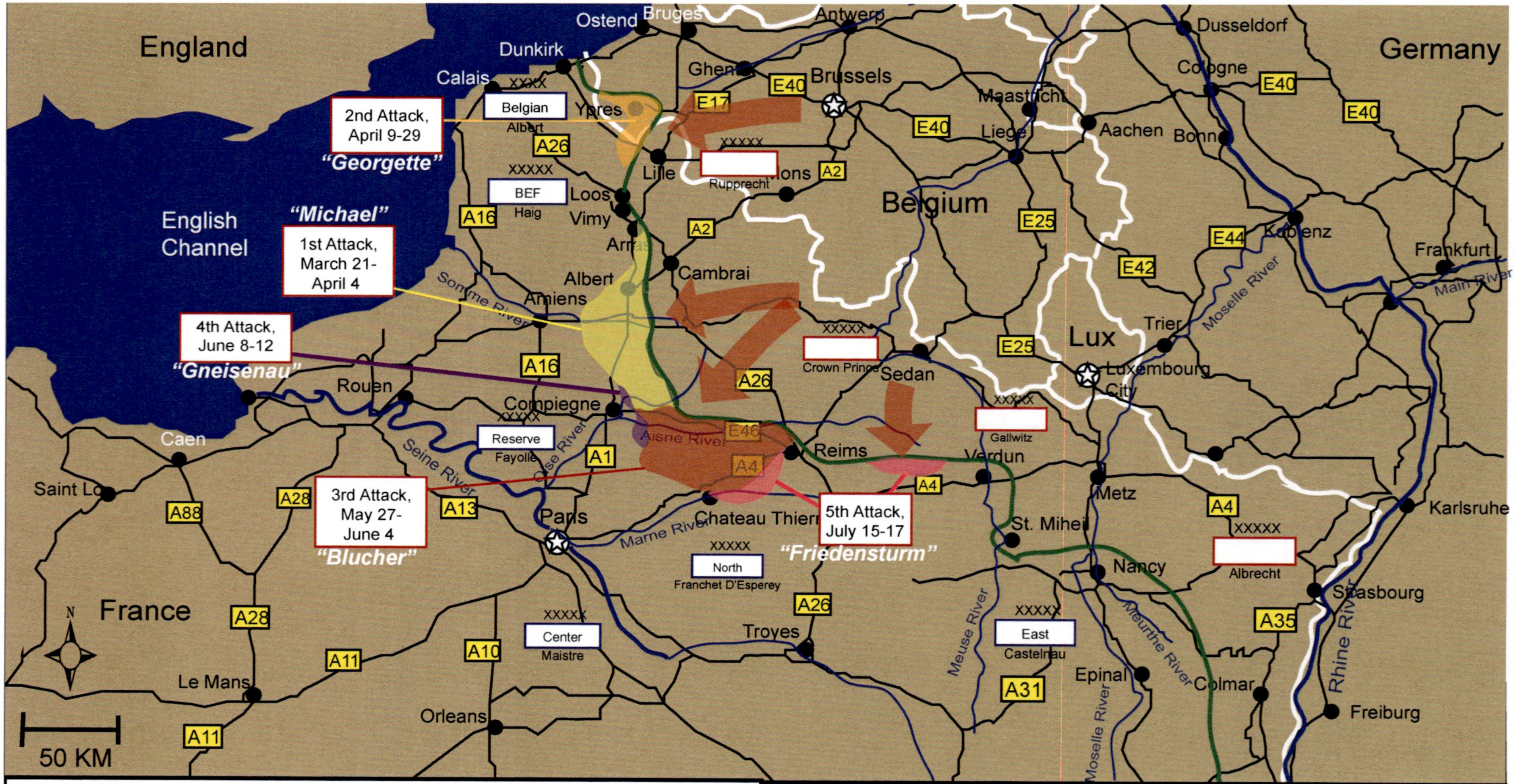

In the Spring of 1918, the Germans wanted to deal a knock out blow on the Allies before the full force of American materiel and men could be used against them. Their first focus was an attack in the north to destroy the British then force the French into an armistice. They did have a big advantage at that time after they had defeated the Russians in the east which freed up 50 more divisions. But after initial gains, the Germans were unable to supply their advances and suffered devasting losses. By July the tide had turned and the Germans could only weakly defended the torn up terroritory they captured. The Allies counterattacked in August and broke through the German lines which led to the end of World War I on November 11, 1918.

Key:
- Situation March 20 1918
- 1st Attack, Somme Offensive, March 21 - April 4
- 2nd Attack Lys Offensive April 9-29
- 3rd Attack, Aisne Offensive, May 27 - June 4
- 4th Attack, Noyon-Montdidier Offensive, June 8-12
- 5th Attack, Champagne-Marne Offensive, July 15-17

BATTLE OF VILLERS – BRETONNEUX

April 24-27, 1918

"Perhaps the greatest individual feat of the war."
—British Colonel George Grogan

SUMMARY: From the end of March, 1918 to April 5, the Germans launched their first Spring Offensive, Operation Michael. The initial attack was formidable and was aimed at Amiens and the British 5th Army. Three German armies pushed back the 5th Army, but the Allied soldiers dug in and stopped the Germans just under 20 km from Amiens. Australian divisions that were part of the British 5th played a crucial role in the sector around the small town of Villers-Bretonneux.

The first Battle of Villers-Bretonneux took place when the Australians stopped the German progress on April 5 on their drive to Amiens. Shortly after, the German command halted Operation Michael. Following a short lull and time to resupply, reposition artillery, the Germans attacked again on April 24, initiating the "Second Battle of Villers Bretonneux."

That battle began with a German artillery bombardment at 0345 on April 24. The aim again was to capture the town of Amiens, an important logistics hub, but the effort also was an attempt to roll up the British front and split it from the French. The British defended Villers-Bretonneux with the veteran, but weakened 8th ID that had suffered more than 4,000 casualties in the first battle. The 8th had the 5th Australian on its left/northern flank, and the 58th British ID on its right. The Germans attacked with 4 divisions, the 243rd and 228th IDs, the 4th Guards and 77th Reserve Division.

After the initial bombardment, the German ground forces attacked. Of great significance, the Germans also deployed 14 of their 20 new A7V tanks. The A7V was an ugly beast; it had one main 57mm cannon, six machine guns, and a total crew of 18! The name was derived from the German department that designed the tank, "Abteilung 7 Verkehrswesen," (roughly translated as Section 7, Transportation). Similar to the British tanks, it was slow and the armor was good for stopping rifle and machine-gun rounds, but still susceptible to spalling on the inside of tank.

As the German tanks came into view, the British 8th Infantry retreated from the line because it had no defense against tanks, allowing the Germans to occupy Villers-Bretonneux. During the battle south of the town, the first tank-on-tank engagement in history took place. On hearing of the German tank attack, 3 British tanks lumbered forward to engage an A7V. The British tanks that attacked included a male Mark IV armed with two 57mm cannons and four machine guns, and two female Mark IVs, each armed with six machine guns. While artillery, machine-gun, and rifle fire continued, the tanks exchanged fire. The German tank hit the 2 female tanks with its main gun, blasting holes in them and forcing them to retreat. Then, the British male hit the German A7V 3 times, forcing it to make a bad turn and fall on its side on an incline. The crew of 18 poured of the tank and the Mark IV gunned down 9 of them. Two more German A7V tanks arrived on the scene with infantry, but were soon pushed away as the male tank fired on them. It then took aim with grape shot against the infantry, killing many of the German soldiers. Then, 9 British Whippet tanks arrived, also using their machine guns to engage the infantry. The British tanks were then greeted with German artillery fire, knocking out 3 of Whippets and damaging the male Mark IV, causing its crew to abandon it and escape to a trench. The German and British crews returned later to retrieve their respective tanks.

After the initial German victory capturing Villers-Bretonneux, on April 24 and only 12 km from Amiens, British 4th Army Commander Gen. Henry Seymour Rawlinson ordered a counterattack by reserve brigades. At 2200, the Australian 13th Brigade attacked south of Villers, followed by the Australian 15th, which attacked north of the town, using a pincer movement to encircle the town.

During the brutal night fighting, Lt. Clifford Sadlier of the Australian infantry gallantly cleared German machine-gun positions to advance. His Victoria Cross citation reads: "On 24/25 April 1918 at Villers-Bretonneux, France, Lieutenant Sadlier's platoon had to advance through a wood where a strong enemy machine-gun post was causing casualties and preventing the advance. Although he was himself wounded, Lieutenant Sadlier at once collected his bombing section and led them against the machine-guns, killing the crews and capturing two of the guns. By this time his party were all casualties and alone he attacked a third enemy machine-gun with his revolver, killing the crew and taking the gun. In doing so, he was again wounded. The very gallant conduct of this officer was the means of clearing the flank, and allowing the battalion to move forward, thereby saving a most critical situation. His coolness and utter disregard of danger inspired all."

At daybreak, various units attacked into the town and captured more than 470 Germans from the 4th German Guards Division. Soon after, the Australians completed the encirclement of the town, cutting off any resistance. South of Villers on April 26, elements of the British 58th and 8th IDs, and the French Moroccan Division attacked Germans in the Bois de Hangard. The attack was a total failure and the Allies suffered greatly trying to press back the German 19th Division.

OUTCOME: The German offensive was pushed back and its artillery was kept out of range of the all-important Amiens rail yards.

WHO? The Allies had committed more than 7 divisions, the Australians suffered 2,400 casualties, the British 9,520, and the French 3,500. The Germans committed 6 divisions and suffered more than 10,000 casualties.

WHEN? April 24-27, 1918

WHY NEAR AMIENS? The Germans began their Spring Offensive to split the British and French forces, capture the logistics hub of Amiens, and roll up the British line to the channel.

PLACES TO SEE:

1. **AUSTRALIAN NATIONAL MEMORIAL AND CEMETERY, VILLERS-BRETONNEUX:** At this location is a cemetery and memorial, considered the National Australian Memorial on the Western Front. Note, the battle took place on the 3rd anniversary of the Battle of Gallipoli, which shares the date for Anzac Day. The memorial sits on the hill the Australians initially charged from to attack the German line north of Villers-Bretonneux.
   N 49 53.143 E 2 30.513 (49.885715, 2.508555)

2. **FIRST TANK-V-TANK MONUMENT:** There is a small stone on the side of the road memorializing history's first tank versus tank battle. The British tanks came out of the woods to the west to meet the German tank crossing the fields to the east. We really need someone to donate a replica Mark IV and a replica A7V for display here!
   N 49 51.640 E 2 29.796 (49.860671, 2.496605)

3. **MUSEE FRANCO–AUSTRALIAN, VILLERS-BRETONNEUX:** In the town here is an excellent small museum with artifacts, photos, and memorials to the battle.
   N 49 51.971 E 2 31.020 (49.866186, 2.517000)

4. **BRITISH CRUCIFIX CORNER CEMETERY:** This contains the graves of 660 Commonwealth soldiers, 141 French, and 2 Russian. Many of them died during the April, 1918 battle, but some died in the August, 1918 offensive.
   N 49 51.199 E 2 31.360 (49.853311, 2.522668)

5. **BRITISH ADELAIDE CEMETERY, VILLERS-BRETONNEUX:** At this location to the west of Villers-Bretonneux is a Commonwealth cemetery with the graves of 960 warriors.
   N 49 52.215 E 2 29.887 (49.870254, 2.498117)

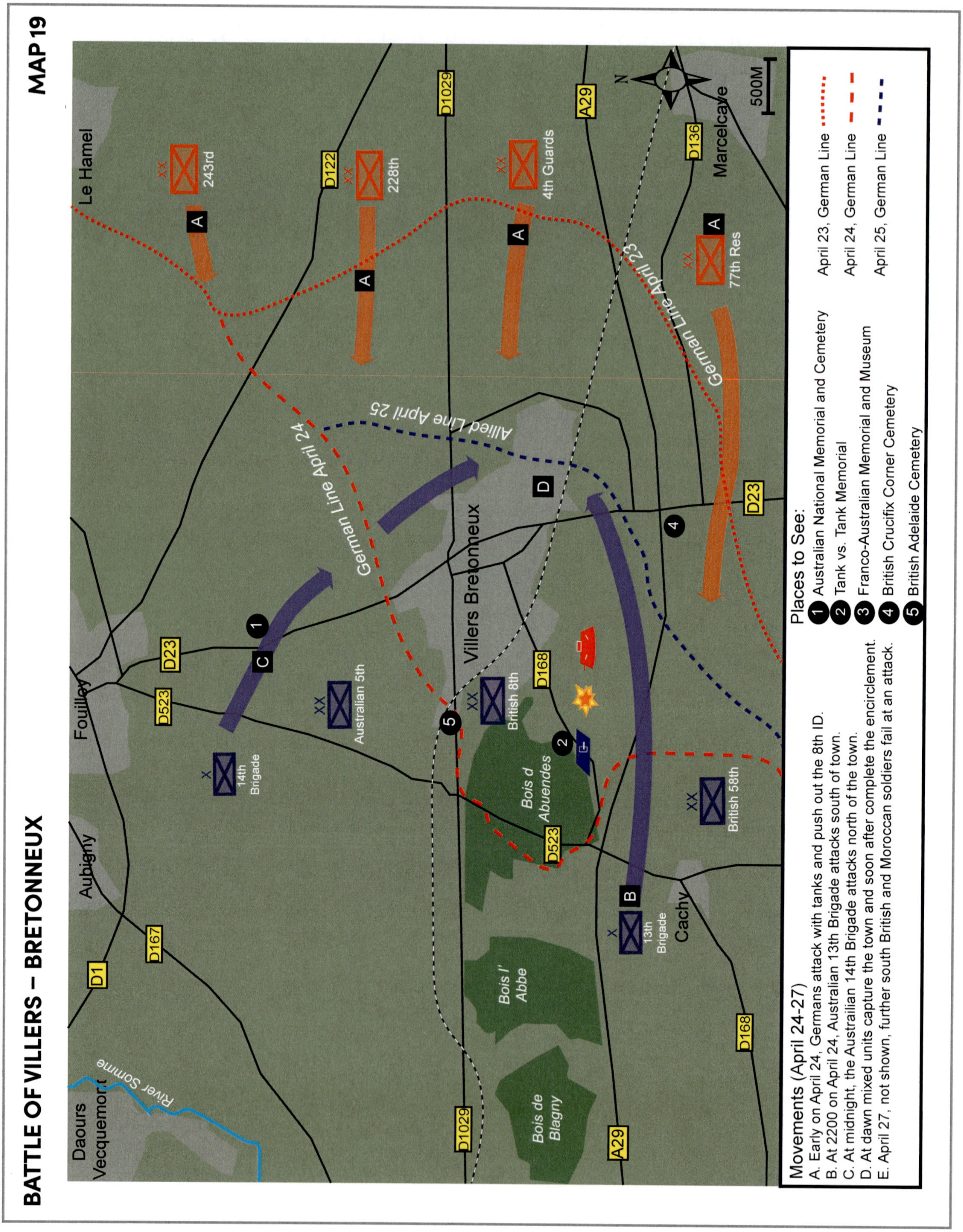
BATTLE OF VILLERS – BRETONNEUX
MAP 19
Le Hamel
Fouilloy
Aubigny
Daours
Vecquemont
River Somme
Villers Bretonneux
Marcelcave
Cachy
Bois d Abuendes
Bois l' Abbe
Bois de Blagny
243rd
228th
4th Guards
77th Res
Australian 5th
British 8th
British 58th
14th Brigade
13th Brigade
German Line April 23
German Line April 24
Allied Line April 25
D1
D167
D523
D23
D122
D1029
A29
D136
D168
500M
N
Movements (April 24-27)
A. Early on April 24, Germans attack with tanks and push out the 8th ID.
B. At 2200 on April 24, Australian 13th Brigade attacks south of town.
C. At midnight, the Australian 14th Brigade attacks north of the town.
D. At dawn mixed units capture the town and soon after complete the encirclement.
E. April 27, not shown, further south British and Moroccan soldiers fail at an attack.
Places to See:
1 Australian National Memorial and Cemetery
2 Tank vs. Tank Memorial
3 Franco-Australian Memorial and Museum
4 British Crucifix Corner Cemetery
5 British Adelaide Cemetery
April 23, German Line
April 24, German Line
April 25, German Line

Photo left is from the front of the Australian National World War I memorial outside of Villers-Bretonneux. It is located just north of the town, from the point where the Australians attacked to encircle the town. The photo right is from the view of the tower looking towards the German line and Hamel is on the horizon to the left in the photo.

An Australian flag at the Hamel memorial and trench lines. On the horizon you can barely make out the Villers-Bretonneux memorial.

Photo left is of the Australians "digging in" near Villers-Bretonneux. (IWM)The photo right is one of the trenches around the memorial at Hamel.

AMERICA GOES TO WAR

Battle of Seicheprey, April 20, 1918
Battle of Cantigny, May 28, 1918
Battle of Belleau Wood, June 3-26, 1918
Battle of Château Thierry, June 3, 1918
Battle of Hamel, July 4, 1918
Second Battle of the Marne, July 18-August 6, 1918
Battle of Fismes, August 4-27, 1918

"Teufel Hunden" (Devil Dog)
—German nickname for the Marines after the Battle of Belleau Wood

---

SUMMARY: After the Americans declared war on Germany on April 6, 1917, the first units of the American Expeditionary Force arrived in France within 30 days. Deployment progress was slow because the American commander, General John J. "Black Jack" Pershing insisted the American forces be well-trained before they went into battle and that they fight as separate units under American command. That preference stood in opposition to the European wish that the Americans serve as individual replacements for European armies. After additional training in France, the US 1st Division entered the defensive line near Nancy in November. When the German offensives took place in the spring of 1918, the Americans had to begin fighting to stem breakthroughs and take part in major counterattacks that ultimately assisted in ending the war, November 1918.

Before discussing the battles, several interesting events and traditions shaped the readiness of the American forces for WW1. During the Civil War (1861-1865) the Army grew to over a million soldiers and the Navy to a few thousand ships. After the war, the suspicious Americans reduced the standing Army to just over 27,000 soldiers and the Navy to 52 vessels, enough to patrol the coast and enough soldiers to fight the Indians. During the reconstruction and industrial boom there were various riots and requirements for policing, the Posse Comitatus Act of 1878 was created and prohibited the Army to be used domestically but allowed the "state militias" to be used for policing, against unions/riots, etc. As a result, this required the "militia" to become more organized and better equipped for the duty of domestic assistance and led to the National Guard. The US was also fortunate for defense to have a corrupt Mexican neighbor to the south, an under populated Canada to the north and oceans to the west and east. These elements allowed the US to maintain a small army and navy.

Shortly after, in the 1880s and 1890s the development of coal power vessels increased the threat of foreign carriers to easily attack the US interests. It was also important for the Navy to protect the US economic doctrine of free trade and open seas. Although the US did not want to colonize lands, it needed bases for coal stations for the steam vessels. Therefore the US had interests in staking primacy in Hawaii, Guam and other islands in the Pacific. Then in 1898 Cuba was in a rebellion against the Spanish colonialists and this impacted the US businesses on the island. Soon after President McKinley deployed naval vessels to protect US interests. That transpired into a war and requirement for the Army and Navy to ramp up capabilities, leadership, organization and equipment to deploy. Several lessons were learned from the poor readiness of the Army for the war. For example, the Army didn't have the professional staff or enough of them to organize recruitment, training, deployment and logistics, the poor sanitary conditions of new bases led to disease outbreaks and there were not enough weapons and uniforms.

Following the Spanish American War, the US built the Great White Fleet to patrol and protect free trade and its new bases; it was a visual reminder of President Theodore Roosevelt's policy to "speak softly and carry a big stick." This foreign policy continued up to WW1, with the US forces intervening in South America several times for "business interests". When the war kicked off in Europe in 1914, the US avoided entry but realized they may have to fight. The movement had support which established in 1915 training camps to develop a system of trained reserves. In 1916, the US passed acts to expand the Naval Fleet and an increase to the Army up to 175,000 men and the National Guard to 400,000 men. Then in 1916-1917, the Mexican Expedition took place. Brigadier General John Pershing was selected to lead the Expedition and this proved successful and allowed for more lessons learned, such as motorized vehicles, supply and command structures.

In November 1916, President Woodrow Wilson was reelected as president and promised to keep the US out of the war. However, in February 1917, the Germans resumed unrestricted U-boat warfare, after a lull in the activity since the German public relations disaster of sinking the Lusitania on May 7, 1915. Then in early 1917, British Intelligence exposed to the Americans a letter from the German Foreign Secretary Arthur Zimmermann to the Mexican government asking for their support to attack America. The letter written in January 1917, authenticity was questioned until Zimmerman confirmed it was true. Shortly after, on April 6 Congress authorized a declaration of war.

Major General Pershing was ordered to France in May 1917, marched in Paris with a battalion of soldiers on July 4, 1917 and then he took command of the American Expeditionary Force (AEF) on July 5. He led the effort to organize the staffing, deployment plan, training, logistics and operations. Filling the ranks, setting up training sites was difficult, let alone the doctrine and equipment to fight. Much of the equipment and uniforms came from the Allies or licenses issued by the Allies for US manufacturers to make such as the British helmet, French tanks and British 1917 Enfield.

Shortly after the Americans arrived in Europe, the Russians surrendered in December, 1917, freeing up 50 German divisions. As described, the following spring of 1918, the Germans launched their Spring Offensive of 5 attacks. Knowing the American forces were building up, the German command wanted to give everything it had quickly to force England out of the war and reach an armistice. The offensives achieved some success; for example they reached Château Thierry in the south, only 50 km from Paris. The US forces rushed into the line to stop the Germans in the French sectors near Château Thierry. The following are summaries of the famous battles in which US forces proved their worth.

APRIL 20, 1918 – BATTLE OF SEICHEPREY: The National Guard, 26th Infantry Division was deployed in defensive positions around the town of Seicheprey, a relevantly quiet area of the front south of Verdun and Metz. Early morning on April 20, the Germans used approximately 3,000 Stormtroopers to attack the American forces to test their resolve. The 102nd Regiment was hit hard and retreated but then counterattacked, to include the support personnel. After hours of close quarter combat, the Germans retreated leaving behind 650 American casualties and taking 100 prisoners. The Germans suffered approximately 100 casualties. It was a hard and sharp fight but the American line held.

MAY 28, 1918 – BATTLE OF CANTIGNY: The Battle of Cantigny marked the first attack by an American division in World War I. The battle was a success, leading to the capture of the town of Cantigny and a turning point against the Germans' Spring Offensives. The 1st Division, 28th Regiment was selected for the attack, along with elements from the 18th Regiment and some French support. To prepare they trained with a full sized mockup of the village. On May 28, starting circa 0500 after an hour artillery barrage the infantry attacked supported by 12 tanks and French aircraft. The assault was major success, after fierce fighting, they took the town in 45 minutes.

After capturing the town, the Americans were hammered by fierce German counterattacks, but repulsed all of them. The American regiment of 3,564 men suffered 1,067 casualties, including 199 killed in action; German suffered approximately 1,500. As a result of their success, the Allies continued with a counteroffensive that gained back land lost earlier in the German offensives.

The 1st Division remained in the line for 71 days and suffered more than 5,200 casualties. Pvt. Daniel Edwards was awarded the Distinguished Service Cross for his actions during Cantigny. His citation reads: "While serving with Company C, 3d Machine-Gun Battalion, 1st Division, A.E.F., at Cantigny, France, May 28 - 30, 1918. Serving as gunner of his machine-gun squad Private Edwards advanced with the first assault line of the Infantry and while passing through the village of Cantigny at 5:30 a.m., May 28 carrying his machine gun upon his shoulder, he was attacked by an enemy soldier and bayoneted, receiving a severe wrist wound; the enemy soldier was killed by an infantryman. Continuing in the advance beyond Cantigny and meeting intense enemy fire, the attacking wave was halted. Private Edwards with his squad remained in an advanced position, protecting with his fire the Infantry which had fallen back to a more advantageous position and were entrenching. While thus engaged the machine gunners repulsed two determined enemy counterattacks, during which the three members of the squad accompanying Private Edwards were killed and he himself severely wounded. Despite these wounds, which he himself dressed, he remained alone in his position throughout the day, firing whenever a target offered, withstanding attacks by liquid fire and machine-gun fire, he refused to be evacuated and continued to operate his gun until nightfall, when his company was relieved. His extraordinary bravery and devotion to duty, his fortitude and undaunted determination despite his numerous and painful wounds, incited the men of his battalion to splendid endeavors and raised their morale to an extremely high pitch."

JUNE 3, 1918 – BATTLE OF CHÂTEAU THIERRY: On May 27, the German's commenced their third spring offensive, Operation Blucher between Soissons and Reims. The German attack blew through the British and French lines, they crossed the Aisne River and Vesle River in under 24 hrs and within the first three days captured 50,000 Allies soldiers. The French quickly requested the Americans fill the gap. General Pershing sent his only available operational units, the 2nd and 3rd IDs. By May 31 the Germans reached the northern bank of the Marne River at Chateau Thierry, only 100 KM from Paris.

There on June 1, the 3rd ID, "the Rock of the Marne," halted the German attack over the Marne River, knocking units back across the river after costly fighting. That same day the 3rd ID blew the main bridge over the river which forced the Germans to pivot west. The Germans then captured Vaux, Belleau Wood and Chateau Thierry on the northern bank. At Belleau Wood, the 2nd ID arrived from the west. The 2nd ID was made up of a brigade of regular infantry, a machine-gun battalion, and a brigade of Marines.

Late at night on June 1, when the Germans pivoted to the west, they broke through a French unit on the left flank of the Americans. Elements of the 23rd Regiment, 6th Machine Gun Battalion and the 1st Battalion 5th Marines forced marched 9 KM to fill the gap. The following day, June 2, the Germans attacked west, to the objectives of Lucy-le Bocage and Marigny en Orxois. The French asked the Marine brigade commander, James Harbord, to retreat. He ignored the request and said, "Hold where we stand." The Marines dug in and mowed down the surprised Germans advancing across the wheat fields.

The French asked again for a retreat and Capt. Lloyd Williams of the 2nd Battalion/5th Marines famously said, "Retreat? Hell, we just got here!" Blocked on the west by the 2nd ID and to the south by the 3rd ID, the German offensive sputtered out and they dug in under the shelter of Belleau Wood, Vaux, Hill 214 and Chateau Thierry.

JUNE 3-26, 1918 – BATTLE OF BELLEAU WOOD: After the Germans stopped their main effort, the Americans were then given the task of clear Belleau Wood, to prepare for a large counter offensive. The match-up consisted of the equivalent of 2 US divisions, with one brigade of Marines, against 5 German divisions.

On June 6 at 0345, the Allies counterattacked. The French 167th Division attacked on the left, the Marines attacked Hill 142 and elements of the 2nd ID attacked Belleau Wood. The Americans had failed to scout properly and did not know that a German regiment had built machine-gun nests throughout their attack zone. Despite the poor reconnaissance, the Americans captured the hill by mid-day, but the battalion suffered 325 enlisted casualties and the death of 9 officers. During the attack on Hill 142, Marine gunnery Sgt. Charles F. Hoffman was awarded the Medal of Honor for extraordinary heroism while serving with the Marine 49th Company, 5th Regiment of the 2nd Division. The first Marine to be awarded the Medal of Honor. His citation reads: "Immediately after the company to which he belonged had reached its objective on Hill 142, several hostile counterattacks were launched against the line before the new position had been consolidated. Gunnery Sergeant Hoffman was attempting to organize a position on the northern slope of the hill when he saw 12 of the enemy, armed with five light machineguns, crawling toward his group. Giving the alarm, he rushed the hostile detachment, bayoneted the two leaders, and forced the others to flee, abandoning their guns. His quick action, initiative, and courage drove the enemy from a position from which they could have swept the hill with machinegun fire and forced the withdrawal of our troops."

At 1700 on June 6, the 3rd Battalion, 5th Marines and 3rd Battalion, 6th Marines advanced into Belleau Wood, crossing a wheat field under devastating machine-gun fire. A 2-time Medal of Honor recipient, Dan Daly, rallied his men at the beginning of the attack under heavy fire by saying, "Come on, you sons of bitches, do you want to live forever?" On reaching the German lines, heavy hand-to-hand combat ensued but they gained a foothold in the southern edge of the woods. The foothold was costly, however, the Marines suffered their worst casualties in a day, 31 officers and 1,056 enlisted men.

On June 7, the Germans counterattacked but they were quickly halted. In response, the Marines counterattacked on June 8 and they were also stopped. To try and break the deadlock on June 9, the U.S. and French artillery barrage destroyed the woods. The Germans responded with high-explosive artillery and mustard gas.

Eventually, after six attacks with rifles, bayonets, clubs, and fists through interlocking bands of machine-gun fire, the Marines pushed the Germans out of Belleau Wood. On June 26, the report by 3rd Battalion 5th Marines commander Major Maurice Shearer simply stated, "Woods now US Marine Corps entirely." The action in Belleau Wood earned the Marines the German nickname "Teufel Hunden" because of their courage and tenacity in the fight. The Americans emerged from Belleau Wood with 7,966 wounded and 1,811 killed. The Germans suffered unknown casualties. The sacrifices showed American skill, resolve and gave a much needed boost to the British and French.

JULY 4, 1918 – BATTLE OF HAMEL: The Battle of Hamel was an Australian Army operation using 5 Australian divisions, supplemented with 4 companies of American infantry. The battle is a highlight of the effectiveness of combined arms and thorough planning. The town of Hamel and surrounding hills provided fields of fire for the Germans on the Allied lines and it was critical that it be captured. It is located south of Albert and only 22 km east of Amiens. The Germans had held the position since the Spring Offensives on April 4, 1918.

To prepare for the battle, the Australian commander, Lt. General John Monash, briefed all levels of troops on their objectives and trained them together, including the supporting tanks. For the main attack, Monash had the 4th Brigade of the 4th Australian ID and 11th Brigade of the 3rd Australian Division, along with armor, artillery, and aircraft. The Hamel defense was held by the German 13th ID concentrated on strongpoints at the villages of Hamel and Vaire on the Somme River.

On the day of the battle, the Allies opened up with 600 artillery pieces at circa 0315, providing smoke and creeping barrage for the infantry. The infantry advanced 100 yards behind the barrage with 60 tanks, 4 dedicated to resupply. Aircraft provided support by dropping resupplies at predesignated areas or by signal flare, and they strafed and bombed German positions and reserve areas. Incredibly, in just under 95 minutes, the units captured all their objectives. As expected, the Germans counterattacked in the early morning and again late that night, but failed to advance against the strongly defended positions of the Australians. After the battle, an American soldier, Cpl. Thomas Pope, was awarded the Medal of Honor for his actions. His citation reads: "His company was advancing behind the tanks when it was halted by hostile machinegun fire. Going forward alone, he rushed a machinegun nest, killed several of the crew with his bayonet, and, standing astride his gun, held off the others until reinforcements arrived and captured them."

In addition, the Australian soldier Henry Dalziel was awarded the Victoria Cross for his actions. His citation reads: "For most conspicuous bravery and devotion to duty when in action with a Lewis gun section. His company met with determined resistance from a strong point which was strongly garrisoned, manned by numerous machineguns and, undamaged by our artillery fire, was also protected by strong wire entanglements. A heavy concentration of machine-gun fire caused many casualties, and held up our advance. His Lewis gun having come into action and silenced enemy guns in one direction, an enemy gun opened fire from another direction. Private Dalziel dashed at it and with his revolver, killed or captured the entire crew and gun, and allowed our advance to continue. He was severely wounded in the hand, but carried on and took part in the capture of the final objective. He twice went over open ground under heavy enemy artillery and machine-gun fire to secure ammunition, and though suffering from considerable loss of blood, he filled magazines and served his gun until severely wounded through the head. His magnificent bravery and devotion to duty was an inspiring example to all his comrades and his dash and unselfish courage at a critical time undoubtedly saved many lives and turned what would have been a serious check into a splendid success." The battle was a great success against the Hindenburg Line, showcasing effective combined-arms tactics. The advance allowed the Allies to position themselves well for additional attacks the following month.

JULY 18-AUGUST 6, 1918 – SECOND BATTLE OF THE MARNE: At 0415 on July 15, along an 80 KM front with three armies the Germans commenced their fifth and final offensive, Operation Marneschutz. The French defense held east of Reims but relied back west of Reims. The French had also developed the defensive doctrine of elastic defense, to allow the German war machine to advance and eventually wear itself out, then counterattack. The French were also made aware of the attack from prisoners and fired artillery on the gathering German troops on July 14-15. Despite this, the

Germans had crossed the Marne River at Dormans and began to threaten Reims by July 17. But the determined French counterattacked in several areas and their artillery disrupted the German supply network. By nightfall on July 17, the Germans were again unable to move forward.

In response to the German attacks, the Supreme Commander Ferdinand Foch had planned a surprise counterattack, the Allies' first major offensive in a year. The goal was to cut off the German salient between Soissons and Reims. On July 18 at 0445 hours the Allied artillery opened up and 16 divisions attacked the bulge from the west. The 6th and 10th French Armies (containing 5 American divisions) attacked on a 40-km front, catching the Germans off guard. In the first day, it is estimated they captured over 10,000 Germans and 250 cannons.

Two days later on July 20, the French 9th and 5th Armies attacked from the south over the Marne River. The attack was supported again with British and American divisions. The Americans captured their objective, Hill 204 outside Chateau Thierry, and took Montair by 1030. One American soldier, Sgt. J.F. Brown, and 12 of his fellow soldiers held out in pocketed positions and returned with battle plans and 155 prisoners. The Germans fought tenaciously as they regrouped and retreated, especially in the center over the heights of the Ourcq River near Fere en Tardonois. The Allies continued to push the Germans back until August 6, where the Germans set up their defensive line at the Vesle River. Both sides were exhausted and the losses were devastating for both sides. The French suffered 95,000 casualties, the British 13,000 and Americans 12,000. The Germans suffered over 168,000. The extensive territory gained, however, was a major turning point for the Allies

AUGUST 4-27, 1918 - BATTLE OF FISMES - FISMETTE: During the 2nd Battle of the Marne, the American 32nd ID, operating within the French 6th Army, drove the Germans back across the Vesle River and captured the town of Fismes on August 4. The Allied forces just slugged it out for 3 weeks, pushing the Germans back 32 km. Fismes is located on the southern side of the Vesle River, Fismette the northern side. Historically, the town has been a critical crossing point of the river and been fought over during several wars. For instance, Napoleon was headquartered briefly in Fismes in 1814 during his retreat, and it was fought over in the Franco-Prussian War. During WW1, the bridge was destroyed 3 times and again in 1940 during WW2.

After the 2nd Battle of the Marne, the 32nd ID was replaced by the Pennsylvania 28th ID on August 5. On August 6, elements of the 112th Regiment of the 28th ID were ordered by the French to capture and hold Fismette across the river. The fight was fearsome; the Germans held the slopes around Fismette, allowing them to use artillery, machine-gun, and sniper fire on anyone who crossed the foot bridge or the remains of the old bridge. That said, despite the German advantages the Americans captured a portion of the town and held.

The close-quarter combat earned several citations for bravery, including to Pvt. James I. Mestrovich, who was awarded the Medal of Honor. His citation reads: "Seeing his company commander lying wounded 30 yards in front of the line after his company had withdrawn to a sheltered position behind a stone wall, Sergeant Mestrovitch voluntarily left cover and crawled through heavy machine gun and shell fire to where the officer lay. He took the officer upon his back and crawled to a place of safety, where he administered first-aid treatment, his exceptional heroism saving the officer's life." In addition, 2nd Lt. Josephy Landry, from company B of the 112th Infantry Regiment, courageously stopped the first German attack on August 27 and used two 1911 .45 pistols, one in each hand, to defend his position. Before dying of a sniper wound, in total during the battle he single-handedly killed 44 Germans.

The Germans, however, continued firing mustard gas and artillery on the American holdouts. On August 10, a German prisoner advised the Americans that the Germans were building up for an attack by stormtroopers with flamethrowers. Before that, however, the Americans were ordered to attack up the draw to clear the Germans from the hills, which predictably ended in total disaster.

After the Americans pulled back from the foot of the hills on August 13, there was silence. The silence was short-lived, though, and was followed by a fierce stormtrooper attack the prisoner had described. Astonishingly, the Americans still held, through gas, flames, shrapnel, and shell shock. Companies were down to fewer than 50 fighting men. That same night, the 112th was replaced by the 109th Regiment. Fighting continued, and on August 22, the Americans incredibly held the entire town, but the Germans still held the high ground.

On August 26, the American 28th ID commander ordered evacuation of Fismette, but was counter-ordered by the French. That same day, the 112th Regiment sent 236 reinforcements of companies C, G, and H back to Fismette. Smelling blood on August 27, the Germans destroyed the crossings with artillery fire, sealing off any chance for retreat. For 20 minutes, the Germans shelled the Americans' positions, then followed it up with 1,000 stormtroopers with machine pistols, grenades, machine guns, and flamethrowers. The Americans held for hours until they were overrun. That day, they suffered 75 KIA and 127 captured. Thirty swam across the river to escape. The American commander was later scolded by General Pershing for not disobeying the French Commander's orders; the French commander officially apologized for his error in judgment.

Outcome of American involvement: The entry of the Americans into the war, starting at Cantigny, reversed Germany's success. By August 6, the Germans were retreating to their previous lines of defense and had canceled future planned offenses.

TECHNOLOGY/TACTICS: Combined arms began to improve greatly, resulting in increased movement and gains in land. Overall, the change in tactics sped the war's conclusion. The Allies succeeded in using tanks with infantry support to take hardened positions. Also, soldiers began receiving more automatic weapons, such as the Browning Automatic Rifle (B.A.R) to increase organic firepower.

PLACES TO SEE:

THE BATTLE OF SEICHEPREY:

1. **MEMORIAL FOUNTAIN TO THE 26TH INFANTRY DIVISION:** In 1922, the Americans raised $500 in funding to build a memorial fountain to the 102nd Regiment of the 26th "Yankee" Division who defended and fought at Seicheprey on April 20, 1918. At this location is the town church and memorial.
N 48 52.179 E 5 47.438 (48.869644, 5.790634)

THE BATTLE OF CANTIGNY:

1. **CANTIGNY MEMORIAL, AMERICA'S FIRST VICTORY:** The Americans' first offensive battle inflicted over 1,000 casualties on the Germans and captured the German held town. It was a turning point for the Allies and boosted America's hope for success in WW1.
N 49 39.793 E 2 29.456 (49.663217, 2.490933)

2. **CANTIGNY 1ST INFANTRY DIVISION MEMORIAL:** This is located just outside town, dedicated to the American 1st ID that fought at Cantigny.
N 49 39.558 E 2 30.085 (49.6593, 2.501417)

3. **US 4TH DIVISION OBELISK ROAD TO SOISSONS:** This memorial is dedicated to the 4th ID that was part of the attack to drive back the Germans during the summer of 1918.
N 49 18.642 E 3 38.840 (49.3107, 3.647333)

4. **AMIENS CATHEDRAL, PLAQUE TO US SACRIFICES, 6TH REGIMENT, ENGINEERS:** On one of the pillars to the right of the altar in the magnificent Cathedral of Amiens is an engraving that memorializes US soldiers who defended Amiens in March, 1918. The cathedral itself is incredible to see. This coordinate is for the parking lot behind the cathedral.
N 49 53.638 E 2 18.177 (49.893961, 2.302955)

THE BATTLE OF BELLEAU WOOD:

1. **AISNE-MARNE AMERICAN CEMETERY AND BELLEAU WOOD MEMORIAL:** A very impressive cemetery and monument to the 2,288 men who died in the regional fighting. The chapel also contains the names of 1,060 men who have no known grave.
N 49 04.680E 3 17.496 (49.078, 3.2916)

2. **STONE HUNTING LODGE AT BELLEAU WOOD (GERMAN BATTALION HEADQUARTERS):** Just above the cemetery are the ruins of the old German battalion headquarters; you can still see foxholes and trenches. The location was an old hunting lodge. A hike into the woods reveals many trenches, barbed wire, foxholes, and craters.
N 49 04.610 E 3 17.454 (49.076833, 3.2909)

3. **MARINE MONUMENT OF BELLEAU WOOD:** This is just a little farther past the hunting lodge above the cemetery. Artillery pieces from the battle are on display, as are plaques placed by various Marine units that have visited the site.
N 49 04.420 E 3 17.438 (49.073667, 3.290633)

4. **HILL 142:** Located just west of the cemetery, this hill was a bastion of the German defense. On June 6, the Marine 5th Regiment crossed a field under machine-gun fire and charged the hill. The Marines fought fiercely and won the day.
N 49 04.402 E 3 15.575 (49.073367, 3.259583)

5. **GERMAN CEMETERY BELLEAU:** The cemetery holds 8,630 German dead and a mass grave of more than 4,000 Germans. It is located between Hill 142 and the American Cemetery.
N 49 04.974 E 3 17.013 (49.0829, 3.28355)

6. **AMERICAN 26TH DIVISION MEMORIAL CHAPEL:** After July 15, the 26th ID took over this sector of the front and launched its counterattack. Just across the street from the cemetery is a chapel rebuilt by veterans of the 26th Division. The interior contains memorials to various men and units, and includes beautiful stained-glass windows. If the chapel is closed, ask the cemetery's keeper, who has a key.
N 49 04.903 E 3 17.601 (49.081717, 3.29335)

7. **BULLDOG FOUNTAIN, CHÂTEAU STABLES:** This is a famous fountain that quenched the thirst of Army and Marine warriors after the fierce fighting for Belleau Wood. Water was scarce and this fountain gave them much-needed reprieve. It had always had a bulldog emblem on it, but it played to the nickname the Marines had just received as Devil Dogs. Since the end of WW1, it has become part of the pilgrimage for the Marines. Legend says it lengthens the lives of the Marines who drink from it.
N 49 05.019 E 3 17.475 (49.08365, 3.29125)

THE BATTLE OF CHATEAU THIERRY AND THE SECOND BATTLE OF THE MARNE:

8. **CHÂTEAU THIERRY AMERICAN MONUMENT, HILL 204:** A very impressive monument to the American victory above the town of Château Thierry overlooking the Marne River. It is in a beautiful location and was built to commemorate the defensive actions of the Americans who stopped the Germans here. It is the location of Hill 204, also fiercely fought over during the June and July operations.
N 49 02.528 E 3 22.286 (49.042133, 3.371433)

9. **OISE-AISNE AMERICAN CEMETERY AND MEMORIAL:** A quiet and beautiful monument and cemetery that holds the graves of 6,012 American soldiers. The cemetery is located in the center of the location of an American advance into the German front line overlooking the Ourcq River. It also contains the graves of 90 American soldiers who were dishonorably discharged, then executed for rape or murder during WW2.
N 49 12.138 E 3 32.935 (49.2023, 3.548917)

10. **FRENCH MONUMENT, LANDOWSKI'S FANTOMES, BUTTE DE CHALMONT:** A haunting memorial to the battle that took place in the fields. Its phantom statues represent ghosts rising from the battlefield. They include a young recruit, a sapper, a machine-gunner, a grenadier, a colonel, an infantry soldier, an aviator, and a martyr, all escaping from the tomb. The location provides a great panorama of the battlefield where the Allies launched their attacks to recapture ground between July 15 and early August, 1918.
N 49 12.850 E 3 24.548 (49.214167, 3.409133)

11. **MEMORIAL DES BATAILLES DE LA MARNE (DORMANS):** This is a very beautiful chapel on the south side of the Marne, overlooking the site where the Germans and French clashed. It is located at the right flank of the French forces.
N 49 04.302 E 3 38.764 (49.0717, 3.646067)

12. **BRITISH MEMORIAL, SOISSONS:** This memorial is dedicated to the 3,887 British soldiers who died in the area during the German Spring Offensive.
N 49 22.867 E 3 19.747 (49.381117, 3.329117)

13. **FRENCH MEMORIAL, SOISSONS:** Near the British memorial and the cathedral, this memorial commemorates the many French who died in the region.
N 49 22.858 E 3 19.627 (49.380967, 3.327117)

13a. **1ST INFANTRY DIVISION MEMORIAL, BUZANCY:** During the second Marne battle, the 1st Infantry Division fought for four days July 18 – 22, 1918 with the French as part of the counterattack to destroy the German salient. A testimony to the level fighting, in just those four days after advancing 11 km, the 1st ID suffered 2,213 KIA and 6,347 WIA. They were relieved by the Scottish 15th ID on July 22, who also have a memorial in the town.

14. **FISMES AND FISMETTE MEMORIAL BRIDGE:** After the war, through donations the state of Pennsylvania designed and provided money for a memorial bridge over the Vesle River. It was completed with 2 lanterns for the dead at one end and statues of peace and agriculture at the other. The 28th ID, a Pennsylvania National Guard Division and historically the oldest unit in the US Army, fought in the bloody battle for Fismette. In 1928, the bridge was completed and commemorated with a large celebration. In 1940, a section was destroyed when the Germans invaded, but restored in the 1950s. Today, you can see plaques from Meadville, PA, the sister city of Fismes. Many of the boys who died were from Company B, 112th Regiment from Meadville. This coordinate is a small parking area on the northern side of the bridge and also the location for No. 15 below.
N 49 18.768 E 3 40.769 (49.312807, 3.679476)

15. **FISMES - FISMETTE BATTLE WALKING PATH AND INFORMATION BOARDS:** The town has memorialized the battle with an impressive walking tour with 9 information boards providing details of the battle and of the actions of the soldiers who fought here. The coordinate is for the first sign, and walking the town

will only take 30 minutes or so to complete, very much worth the time. On the map, I briefly traced where the signs are located and the walking paths.
N 49 18.768 E 3 40.769 (49.312807, 3.679476)

BATTLE OF HAMEL:

1. **BATTLE OF HAMEL AUSTRALIAN MEMORIAL:** At this location is the memorial to the Australian Corps for its outstanding victory at Hamel. It includes maps and information boards explaining the battle. It is located east of Hamel on the German defensive lines. Also, the German ace, the Red Baron, Baron Manfred von Richthofen was shot down on April 21, 1918 near this location.
N 49 53.989 E 2 34.893 (49.899816, 2.581556)

2. **AUSTRALIAN NATIONAL MEMORIAL AND CEMETERY, VILLERS-BRETONNEUX:** At this location is a cemetery and memorial, considered the National Australian Memorial on the Western Front. Note, the battle took place on the 3rd anniversary of the Battle of Gallipoli, which shares the date for Anzac Day. The memorial sits on the hill the Australians initially charged from to attack the German line north of Villers-Bretonneux.
N 49 53.143 E 2 30.513 (49.885715, 2.508555)

3. **LOCATION OF GERMAN STRONG HOLD "PEAR TRENCH":** At this location there is not a memorial, only the farmers' fields. During the battle it was the strong point in the German line at this location called "pear trench." It is worth driving by the area down the small farm roads.
N 49 53.978 E 2 33.045 (49.899634, 2.550755)

BATTLE OF CANTIGNY, MAY 28, 1918

MAP 20

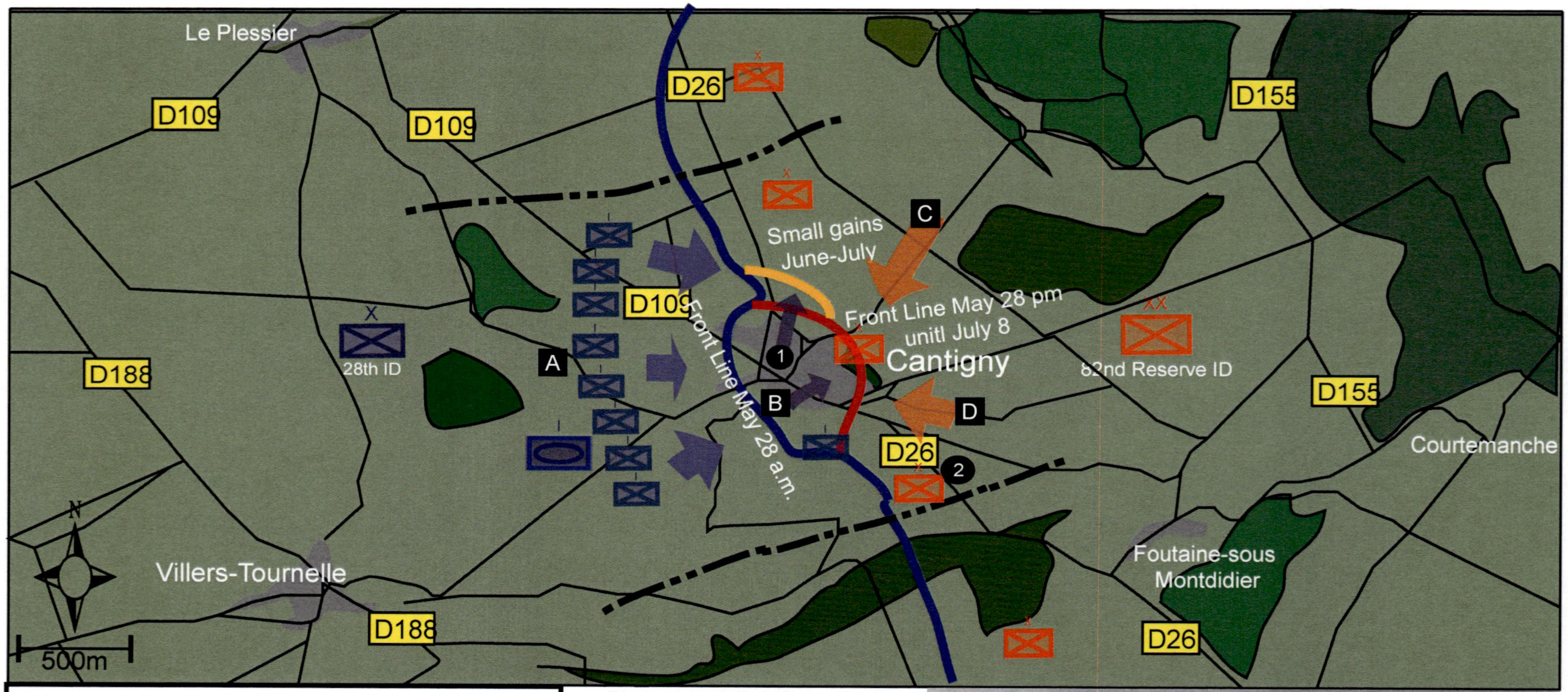

Places to See:

1. American Monument to Cantigny and 50 feet away is the statue dedicated to the 1st Division victory in Cantigny.
2. Memorial to the 1st Infantry Division

Photo right, 28th Infantry Regiment of the 1st ID, go over the top at Cantigny, May 1918.

MOVEMENTS:

A. At 0645 for an hour the American artillery bombarded the German line. Next following a rolling barrage, 8 companies of the 28th Regiment of the American 1st Infantry Division began their advance. Supporting the attack were aircraft and 12 French Schneider Tanks.

B. Within 45 minutes, the Americans captured the town of Cantigny.

C. At 0830 the Germans counterattacked for the first time but were repulsed.

D. Next the Germans attacked again at 1700 and again at 1800 both were repulsed. Selective fighting continued until June 8.

BATTLE OF BELLEAU WOOD, JUNE 3 TO JUNE 26, 1918

MAP 21

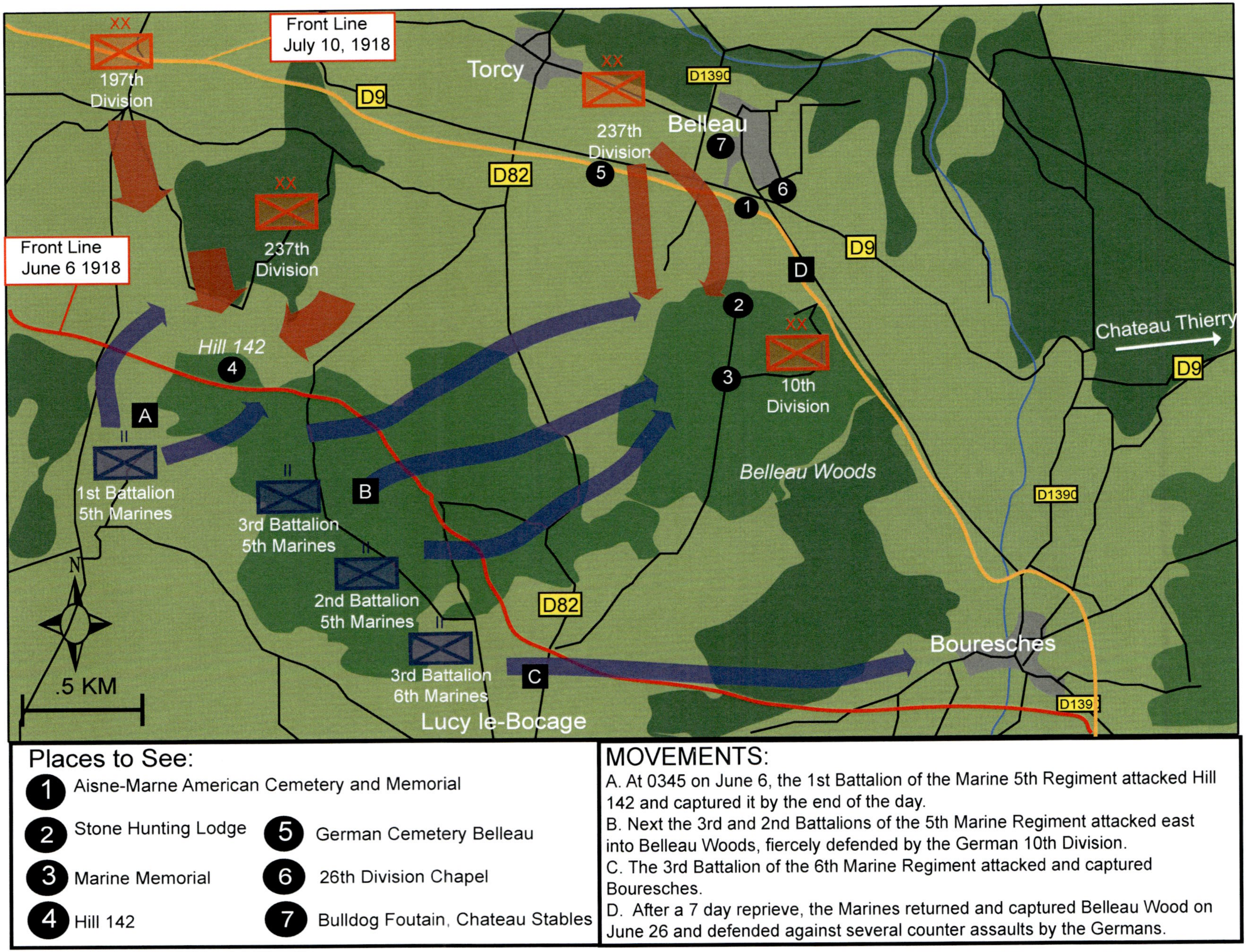

Places to See:

1. Aisne-Marne American Cemetery and Memorial
2. Stone Hunting Lodge
3. Marine Memorial
4. Hill 142
5. German Cemetery Belleau
6. 26th Division Chapel
7. Bulldog Foutain, Chateau Stables

MOVEMENTS:

A. At 0345 on June 6, the 1st Battalion of the Marine 5th Regiment attacked Hill 142 and captured it by the end of the day.

B. Next the 3rd and 2nd Battalions of the 5th Marine Regiment attacked east into Belleau Woods, fiercely defended by the German 10th Division.

C. The 3rd Battalion of the 6th Marine Regiment attacked and captured Bouresches.

D. After a 7 day reprieve, the Marines returned and captured Belleau Wood on June 26 and defended against several counter assaults by the Germans.

BATTLE OF HAMEL, JULY 4, 1918

MAP 22

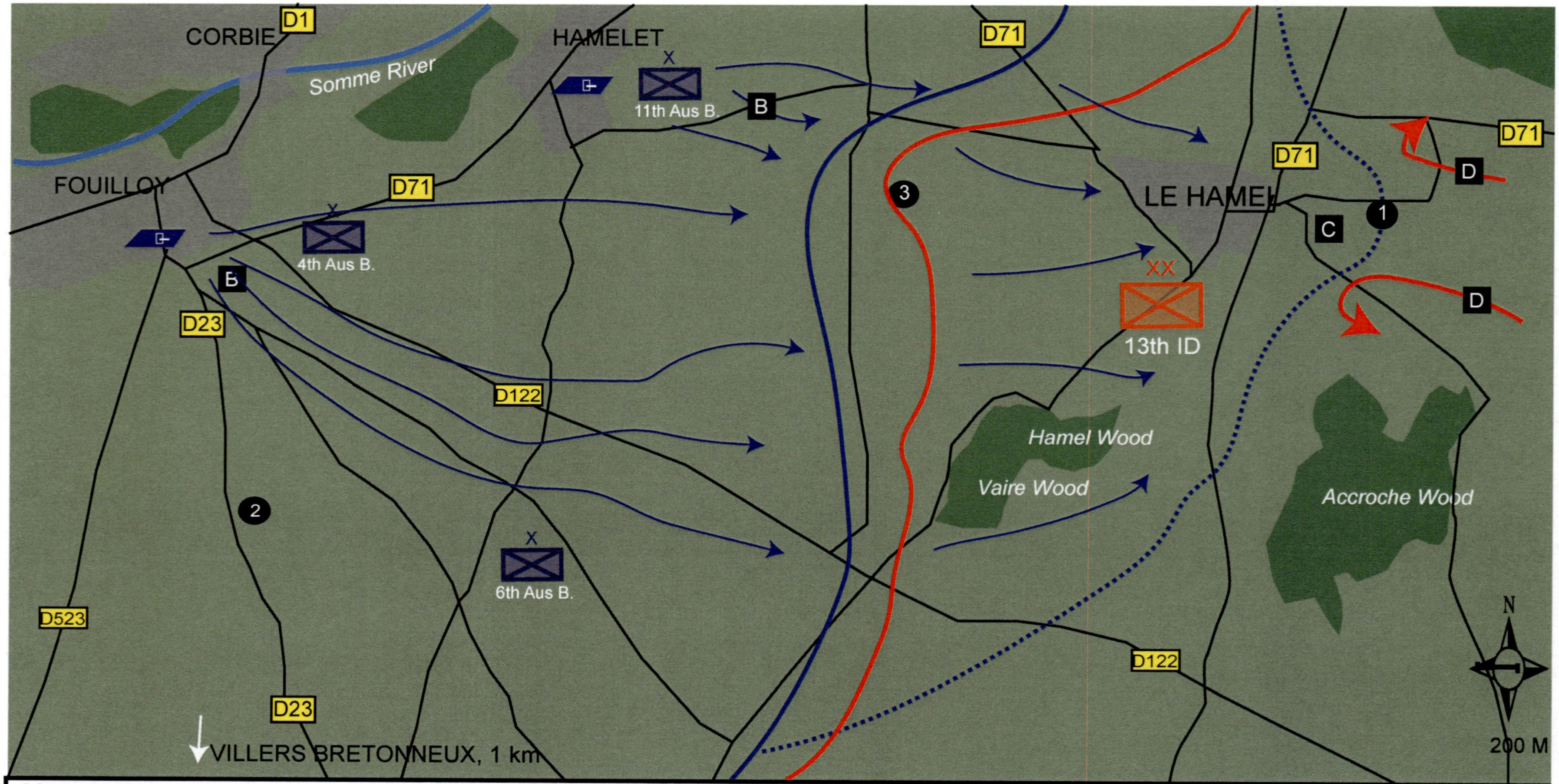

Movements:
A. On July 4, at 0300 the Allied artillery opened fire.
B. Soon after infantry with 60 tanks began to attack the German line
C. In just 95 minutes the Allies captured their objectives.
D. Later in the morning and evening the Germans attempted to counterattack but did not suceed.

Places to See:
1 Hamel, Australian Memorial
2 Villers Bretonneux Memorial and Cemetery
3 Approximate location of Pear Trench

Tank Assembly area
Allied line July 4
German line July 4
Allied line July 4 at 435 am.

MAP 23

SECOND BATTLE OF THE MARNE, JULY 18 TO AUGUST 6, 1918

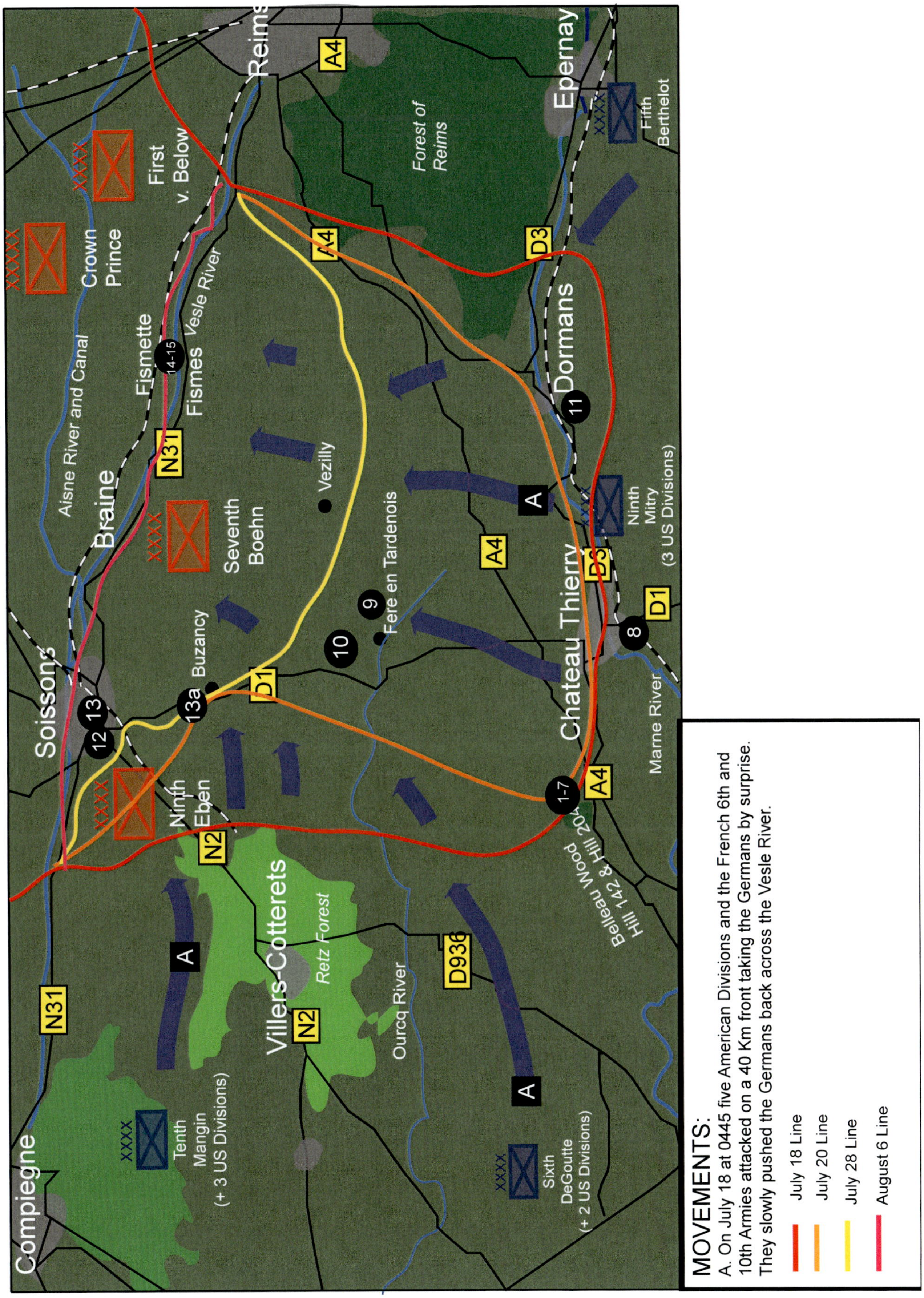

SECOND BATTLE OF THE MARNE, JULY 18 TO AUGUST 6, 1918

MAP 23

PLACES TO SEE

- 1-7 Belleau Wood Places to See
- 8 Hill 204, Chateau Thierry Memorial
- 9 Oise Aisne American Cemetery
- 10 Butte de Chalmont - The Phantoms
- 11 Dormans Chapel and Memorial
- 12 British Memorial to Soissons Battle
- 13 French Memorial to Soissons Battle
- 13a Memorial to American 1st ID, Buzancy
- 14-15 Fismes Memorial Bridge and Path

Above, the Phantoms Sculpture, above right, Hill 204 Memorial and below right, Hill 204 being bombarded by German artillery.

BATTLE FOR FISMETTE, AUGUST 4 TO AUGUST 27, 1918

MAP 24

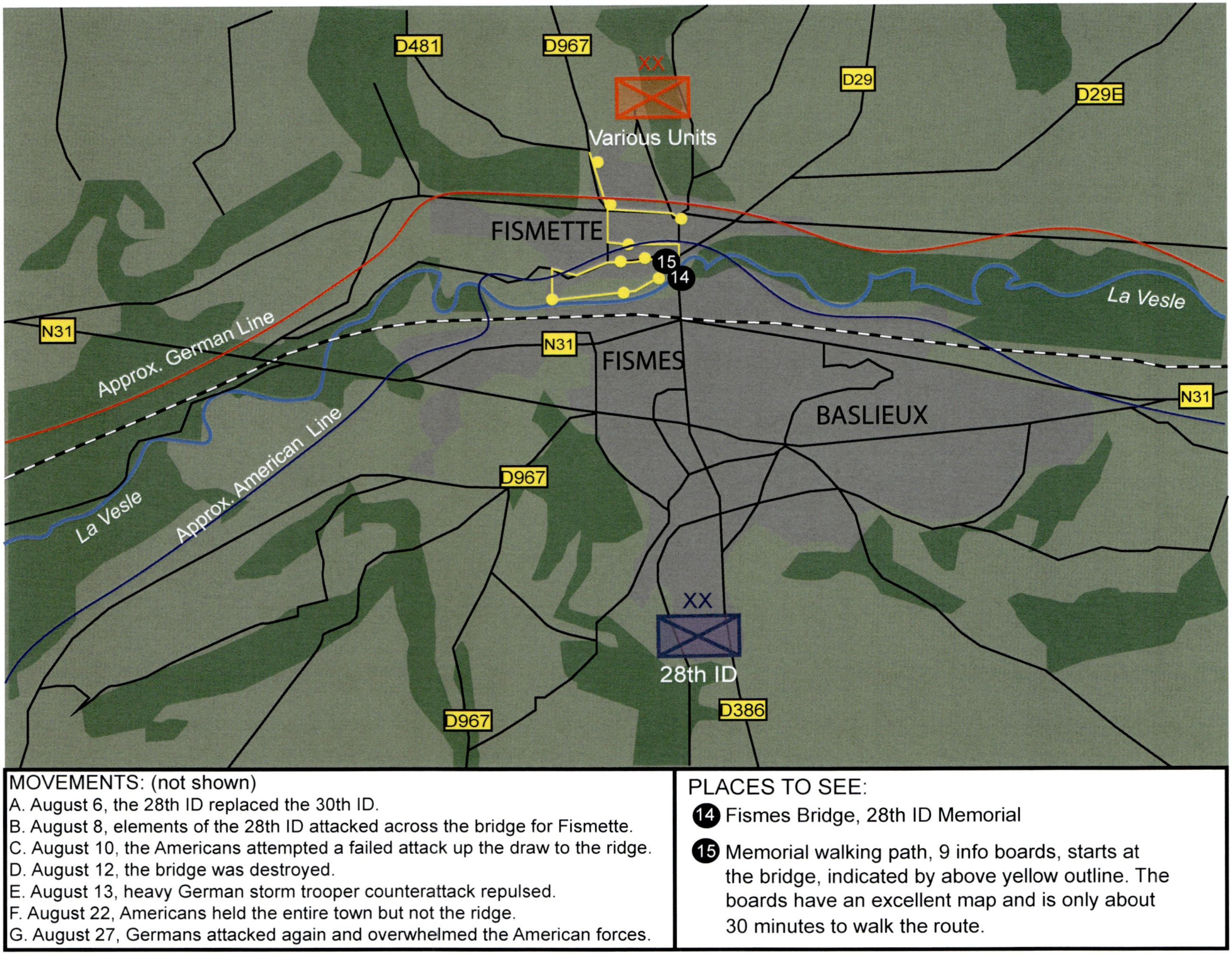

MOVEMENTS: (not shown)
A. August 6, the 28th ID replaced the 30th ID.
B. August 8, elements of the 28th ID attacked across the bridge for Fismette.
C. August 10, the Americans attempted a failed attack up the draw to the ridge.
D. August 12, the bridge was destroyed.
E. August 13, heavy German storm trooper counterattack repulsed.
F. August 22, Americans held the entire town but not the ridge.
G. August 27, Germans attacked again and overwhelmed the American forces.

PLACES TO SEE:
14 Fismes Bridge, 28th ID Memorial
15 Memorial walking path, 9 info boards, starts at the bridge, indicated by above yellow outline. The boards have an excellent map and is only about 30 minutes to walk the route.

Photo left is the Oise Aisne American Cemetery, located only 23 KM NE of Chateau Thierry in a very rural but beautiful area. Photo right is of Dormans, a beautiful chapel and memorial to the French who stopped the German advance.

Photo left, is a view from Hill 204 and the American memorial at Chateau Thierry. The Marne River, is just below. The photo right is one of the many farms on your drive through the fields, this one of laborers picking grapes.

Photo left, the Belleau Wood Marine Memorial surrounded by captured equipment. Photo right, the Belleau Wood American cemetery and memorial. Behind the chapel are the remains of the German Battalion HQ, "the stone hunting lodge."

Far left is the Cantigny memorial to the American soldier. Also nearby is the ABMC Memorial to the Battle of Cantigny.
Below center is the Amiens Cathedral, a breathtaking Cathedral with amazing detail, to include a Gothic era uncle giving nuggies! Amiens was a goal of the German spring offenses.

Above is the Butte de Chalmont, Landowski Phantoms. He started the project in 1919 and unveiled it in 1935. It is located in the middle of the fields in which the Allies launched their July to early August 1918 counter offensive to recapture lost terrority. Above right is a view of the field that many Allied soldiers heroically crossed to defeat the Germans.

The remains of Fismette, looking north over the La Vesle River from Fismes on the southern bank. (NARA)

Below right the photo is from the same side, I believe the yellow house on the left is the same as the one in the photo to the left.
Below left is a plaque from the town of Meadville, PA which many of the 28th ID soldiers came from.

Above is the Fismes town hall, photo right what it looked like after the battle. (NARA) Fismes was also a place of battle during Napoleon's era, his forces marched through here, when out maneuvering the Prussians.

Above a photo of a 28th ID soldier highlighted against a white phosphorous round exploding in Fismette.(NARA)

To the left the river Vesle, about 200 meters down stream, the Americans set up a crossing point to resupply and evacuate troops

To the right, one of the information boards providing excellent info on the soldiers and the events of the battle, along a walking route.

Photo left is from the German point of view looking down onto Fismette and Fismes located across the river. It was a significant advantage to shell the town and any attempts to resupply the towns.

100 DAYS OFFENSIVE

August 8-November 11, 1918

"A great war leaves the country with three armies—an army of cripples, an army of mourners and an army of thieves."—German Proverb

---

SUMMARY: By the end of July, five major German offenses in 1918 had failed to make a decisive breakthrough. In response, Supreme Allied Commander Ferdinand Foch coordinated a final offensive that became known as the "Grand Offensive," or "100 Days Offensive," to breech the Hindenburg Line and defeat Germany. Similar to Ludendorff, Foch planned an offensive with multiple punches across the front to keep the Germans guessing and on their heels. The offensive began with three battles across the front, pushing back the German gains from their Spring Offensives: the Battle of Amiens, the 2nd Battle of the Somme, and the Saint-Mihiel Offensive. These are brief summaries of each major attack during the final months of the Great War.

August 8-11, 1918, Amiens Offensive: The Battle of Amiens punched a 13-km hole in the German line and marked a turning point in the war. Shortly after the Germans' last offensive in July, the Allies quickly regrouped and the British planned a surprise attack. The Allies armed themselves with 1,389 cannon, 10 divisions (150,000 men), 550 tanks, and 1,900 aircraft. The British 4th Army led the attack with the Canadian and Australian Corps at the tip of the spear and the French 1st Army in support. The British attacked north of the Somme River, the Australians south of it. The Canadians were south of the Australians and the French were south of the Canadians. Surprise was the key to success; the Allies kept communication to a minimum, didn't use a pre-bombardment, and moved troops only at night. At 0420 on August 8, the Allies attacked with perfectly timed artillery, advancing 8 km deep and 13 km wide. The Allies suffered 22,200 casualties, the Germans 24,000 and another 50,000 captured. German Commander Erich Ludendorff called it the "Black Day of the German Army." The Germans retreated to the Hindenburg Line with few chances of success, because they had suffered significant casualties and their morale was destroyed. Approximately a week later, the German high command advised the Kaiser that the war could not be won militarily. The Allies had to halt the advance because they had outpaced their artillery and had only 6 tanks battle-ready.

August 21–September 2, 1918, Second Battle of the Somme: After the success of the Battle of Amiens, the British, Americans, Canadians, and Australians attacked the German lines farther north at Albert on August 21. They recaptured Albert on August 22, and by August 29, Bapaume fell. The Australians earned additional distinction on August 31 when they crossed the Somme River and broke through German lines. Then the Canadians gallantly defeated the Germans at the Battle of Canal du Nord, after which the Germans withdrew to the Hindenburg Line, their starting point in the Spring Offensive. Germany had lost all of its gains.

SEPTEMBER 12-18, 1918, SAINT-MIHIEL OFFENSIVE: Before they Allies continued the next phase of the Grand Attack, they drew up plans for the American 1st Army to attack the Saint-Mihiel salient to protect the flank of the upcoming offensive through the Argonne Forest. That was the first battle the Americans would fight under their own command. In August Pershing was gathering his forces to create the Army. Only two days before the battle, Pershing officially activated the 1st Army on September 10. His staff drew up detailed plans for the attack, divisional routes, orders with aerial photos, supply points and intended to maintain Pershing's emphasis for mobile open warfare attacks.

The battle plan included, 3 US corps who controlled 11 US divisions and 1 French division. The French provided 1 corps with 3 French Divisions and Pershing also had three US divisions in reserve. The attack was also supported with the 1st Tank Brigade, commanded by Col. George Patton with 140 tanks and the 1st Assault Artillery Brigade with 270 vehicles. In addition, Col. William Mitchel a father of the Air Power doctrine, controlled over 1,500 aircraft from US, French, British, Portuguese and Italian forces.

They were up against a well-fortified German sector with tunnels and defenses that had been built in an area that had been quiet throughout the previous four years. The Germans only had approximately 10 divisions of soldiers holding the line. The Germans new a build was occurring and expected an attack in 5-7 days and drew up plans to withdraw to The Hindenburg Line; they did not expect it to happen in two days.

The attack started with a four hour artillery barrage on September 12. Mitchel's aircraft over the next 12 hours gained air superiority and began attacking aerodromes, trenches, communication centers and supply depots. The Americans caught the German by surprise in the middle of a planned withdrawal. As such, the first lines were unoccupied and the Americans pushed further to catch the Germans, to include Patton leading an armor spearhead. By September 15, the salient was cleared of the Germans and the battle over. In total both sides suffered approximately 7,000 casualties. The Americans also bagged over 15,000 German soldiers and 450 cannons. Most importantly, the battle was the first Army level battle commanded by the Americans and they performed very well. The combined arms tactics and organization at that level of operations is extremely complex, it gave another confidence boost to the Americans and their Allies.

SEPTEMBER 26-NOVEMBER 11, 1918, CONTINUATION OF THE GRAND OFFENSIVE: After the battles of Amiens, Somme, and St. Mihiel that reduced the recent German gains, Allied Supreme Commander Foch initiated his 3 main thrusts into the German defenses. He meant to disperse any counterattacks and have the thrusts converge on Liege, Belgium. The battles were:

SEPTEMBER 29 – OCTOBER 2, 1918, BATTLE OF ST. QUENTIN CANAL: The Battle of St. Quentin Canal marked the first point where the Hindenburg Line broke, consequently destroying all hope for a German victory. Against 13 German divisions, the Australians, Americans, and British attacked with 14 divisions. After 3 days of battle, the Allied attack broke a 17-km hole in the Hindenburg Line by October 2.

The Australian Corps of the 4th Army was given the task to attack the Hindenburg in the area between Venduile and Bellenglise. The Australian commander, General Monash had difficulty assembling strong enough ranks, to help he was given the American Divisions the 27th ID and the 30th ID for the attack. These divisions were serving under British command in the Ypres battle area. To prepare for the battle, the British began to attack and probe the outskirts of the Hindenburg Line, which was handed off to the 27th ID on September 25. The 27th ID found that the British did not take the attack launching points for the main attack taking place in four days. So elements of the 27th with the help of tanks and artillery attempted to take the strong points of two farms and "the knoll". Over a two day period they too were unable to advance. The actions were intense, to include heroic acts such as Lt. William Turner who was posthumously awarded the Medal of Honor during the skirmishes. His citation reads, "He led a small group of men to the attack, under terrific artillery and machinegun fire, after they had become separated from the rest of the company in the darkness. Single-handed he rushed an enemy machinegun which had suddenly opened fire on his group and killed the crew with his pistol. He then pressed forward to another machinegun post twenty-five yards away and had killed one gunner himself by the time the remainder of his detachment arrived and put the gun out of action. With the utmost bravery he continued to lead his men over three lines of hostile trenches, cleaning up each one as they advanced, regardless of the fact that he had been wounded three times, and killed several of the enemy in hand-to-hand encounters. After his pistol ammunition was exhausted, this gallant officer seized the rifle of a dead soldier, bayoneted several members of a machinegun crew, and shot the other. Upon reaching the fourth line trench, which was his objective, Lieutenant Turner captured it with the nine men remaining in his group and resisted a hostile counterattack until he was finally surrounded and killed."

Despite the skirmish failure, the main attack went forward. On September 29 at 0550 hours in the fog and heavy rain, the Americans were the first to attack, with some tank support. The 27th ID was on the northern/left flank, the 30th ID in the center and the British 46th on the right/southern flank. The Australian 5th Division followed closely behind the 30th ID, to exploit the gaps. During the attack 30th ID, a rather inexperienced force, made progress but failed to mop up the isolated German pockets. Soon after, the Germans came out of deep bunkers along the canal to attack the 30th ID from the rear. The Australians then attacked fighting through a confusing field of Germans and Americans. On the left flank, the 27th ID did not receive effective artillery support from the British against the German hard points, such as "the knoll". As a result, the 27th 107th regiment suffered over 990 casualties in the first day of battle, the most for any American unit during the war in a single day. Despite this, with the help of the supporting Australian 3rd ID, they took the Knoll and strongpoints by 1300 that day. On the southern flank, the British 46th ID was successful in using the fog to cross the canal and gain a foothold on the eastern side. Across the attack front, by the end of the first day, the Allies had punched a hole 6,000 meters deep by a frontage of 10,000 meters, in the strongest portion of the Hindenburg Line. In total the Allies suffered over 12,000 casualties during the four days of battle. The Germans suffered approximately the same with another 6,000 soldiers taken prison.

After the breach, on October 2, the Australians and British reached the Beaurevoir Line (the 3rd row of the Hindenburg Line). On October 10, the Allies had captured the surrounding hills and cleared that portion of the Hindenburg Line. The battle was fierce and deadly, a testimony to the ferocity, 9 Americans were awarded Medal of Honor citations. For example, Sgt. Joseph Adkison of Company C, 119th Infantry, 30th Division, for action on September 29, 1918. His reads: "When murderous machinegun fire at a range of 50 yards had made it impossible for his platoon to advance, and had caused the platoon to take cover Sergeant Adkison alone, with the greatest intrepidity, rushed across the 50 yards of open ground directly into the face of the hostile machinegun kicked the gun from the parapet into the enemy trench, and at the point of the bayonet captured the 3 men manning the gun. The gallantry and quick decision of this soldier enabled the platoon to resume its advance."

SEPTEMBER 26 TO OCTOBER 28, 1918, THE CHAMPAGNE OFFENSIVE: During the last and 5th German offensive on July 15 the majority of forces attacked west of Reims near Chateau Thierry. At the same time, they also attacked July 15 east of Reims. Similar to the western attack it was stopped but there was no major follow on counterattack by the Allies, like to the west. Early in September Foch had ordered the French 4th Army to prepare for an attack on the German line in unison with the Americans to the west in the Argonne, starting September 26. Their objective was to push the Germans back to the Aisne River, while protecting the American left flank who was attacking into the Argonne towards Sedan.

On September 26, after a six hour artillery bombardment, two French corps advanced behind a rolling barrage. Despite the Germans attempting an elastic defense the French still took 2.5 km, recaptured Navarin, Souain and Tahure, all lost in July 1918 to the Germans. Progress was slow but by September 30, the French captured 8,000 POWs and 300 artillery pieces. The Germans in response sent in ten reinforcement divisions, which blunted the French advance, unable to capture Mont Blanc and Hill 210 which is a crescent moon shaped ridge north of Sommepy. To capture the ridge, the French called up the American 2nd ID on October 2. The American commander devised a pincer attack that sent the 4th Marine Brigade to the left and the 3rd Infantry Brigade to the right. On October 3 at 0550 the attack begun after a preparatory artillery barrage. The Germans actually attacked as well for a spoiling attack which was swiftly repulsed by the Americans. By 0830 the 3rd Infantry Brigade captured the eastern slope of the crescent and the strong point of Medeah Farm on the right flank. The Marines slugged their way through the right flank by 0900, capturing the summit of Mont Blanc. The attack was a success but it created a salient which the Germans could pour in fire from three sides. That night, German reinforcements arrived and they pulled up additional machine guns and artillery on the flanks. The next day on October 4 at dawn the German 15th Bavarian Division attacked in their pincer attack from St. Etienne and Medeah farm, both attacks were destroyed by the Americans. On October 5, elements of the 3rd Battalion 6th Marines attacked and captured the remaining hold outs on Mont Blanc, a testimony to the fire power remaining the Marines captured over 200 POWs and 75 machine guns! That same day the French 22nd ID outflanked the German positions around Notre Dame des Champs. They and the Marines then attempted but failed to take St. Etienne.

On October 7, the battle worn 2nd ID was replaced by the American 36th ID. On October 9, the 36th ID with supporting French tanks and infantry captured the town of St. Etienne Arnes. In the following days, the French and Americans pursued the German retreat for a bloody 23 km to the Aisne River. Attempts were made to cross but the German defense was too strong. On October 28, the offensive was halted.

SEPTEMBER 28, 1918, 5TH BATTLE OF YPRES: The 5th Battle of Ypres took place during the 100 Day Offensive and had great success against the demoralized Germans. The battle was led by Belgian King Albert I. Under his command, the Allied divisions gained 29 km and recaptured most of what was lost during the Spring Offensive. They had to stop the advance, however, due to a shortage of supplies.

SEPTEMBER 26, 1918, MEUSE ARGONNE OFFENSIVE: This was the largest commitment of the AEF during the war and one of the bloodiest battles in US military history. The attack was part of Foch's overall strategy after the multiple punches across the front for the last two months. After the success of the St. Mihiel Offensive, Pershing wanted to continue north towards Metz, threaten Sedan and the Metz – Lille rail line supporting the entire German front. Foch wanted the Americans to threaten the line but from the position west of Verdun, closer to the entire allied force advancing. As a result, Pershing had to shift his forces to starting points west of Verdun and build up his resources. Pershing left 1 corps in the St. Mihiel area and assembled the rest of his forces for the attack scheduled only 11 days later. His success relied heavily on another soon to be famous, Col. George C. Marshall, who was in charge of the logistical planning. On September 11, Marshall was given the order to organize the materiel and men for the offensive taking place on September 26. The task required moving 500,000 men, 2,000 cannons and 900,000 tons of supplies/ammo.

For the attack, Pershing deployed three corps, each with three divisions and one in reserve, in addition the Army command had three more divisions in reserve. In total, a force of over 225,000 American soldiers, 2,780 cannons, 380 tanks and 840 aircraft were assembled for the initial attack. The objective was to capture the important rail junction of Sedan, 70 KM north of the starting line. In the American forces way was the difficult terrain of the Argonne Forest and experienced, well entrenched 190,000 German soldiers, close to their own supply. The battle would consist of three phases.

The first phase began on September 26 at 0230 a three hour artillery barrage by 2,700 cannons initiated battle. Unlike the St. Mihiel battle, the Germans were staying to fight, not withdrawing. The American divisions committed also had many with less training and no experience. Initially, the Americans made some gains but the terrain and German resistance began to slow and stop the American advance. The Germans even counterattacked in some locations against the 91st ID who had just broke through two lines. Due to the terrain, American units were cut off from each other and the German reinforcements began to arrive, a total of another 7 divisions. Strategic locations such as the heights of Montfaucon saw vicious combat, there after three days of battle on September 29 it was captured. On October 1, the advance was halted, the French, west of the Americans, had gained 14.5 km in the open terrain, compared with the Americans, who had gained only about 3 km. The American fighters included future

President Harry Truman, then-Col. Douglas MacArthur, and future founder of the Office of Strategic Services, Maj. (later General) William Donovan.

The 2nd phase of the Meuse Argonne Offensive began on October 4, to continue grinding forward towards Sedan, which the Germans recognized the great threat and poured in more reinforcements. The second phase, resulted in several citations for now-famous Americans. The Lost Battalion of the 77th Division received 5 Medals of Honor, including one for Maj. Charles White Whittlesey after his battalion advanced too far into the Argonne Forest ahead of the units on its flanks. After advancing through the German lines, believing Allies were still on their flanks, 554 men found themselves outnumbered and surrounded for 6 days behind German lines. After fighting off several German attacks, only 194 of the battalion's 554 men were able to walk out.

A second famous citation during this battle was the Medal of Honor to Cpl. (later Sgt.) Alvin C. York, of the 82nd Division. On October 8, during a patrol along the Decauville rail line in Chatel-Chéhéry, a York-led 8-man unit killed 28 Germans, took 132 prisoners, and silenced 32 machine guns. York later stated, "A higher power than manpower guided and watched over me and told me what to do." York's citation reads: "After his platoon had suffered heavy casualties and 3 other noncommissioned officers had become casualties, Cpl. York assumed command. Fearlessly leading 7 men, he charged with great daring a machinegun nest which was pouring deadly and incessant fire upon his platoon. In this heroic feat the machinegun nest was taken, together with 4 officers and 128 men and several guns."

Part of the second phase, the Americans launched attacked at Romagne, where they finally broke through the Hindenburg Line on October 14. The Americans began also infiltrating and bypassing the German strongpoints gaining territory at night. By October 28, the AEF had moved about 18 km and cleared the Germans out of the difficult terrain of the Argonne Forest.

The third phase began on November 1, after the French and Americans refreshed and reequipped. The attack began with the French 4th Army on the left, the American 1st Army in the center and the newly activated American 2nd Army driving on the left towards Metz. (Pershing had moved back to AEF HQ in Chaumont to control all the AEF armies.) Soon after, the Americans captured the a German fort allowing the French to cross the Aisne River and enter Sedan on November 6, which had been a possession of Germany since they captured Emperor Napoleon III in 1870. The American 2nd Army was close to capturing Metz but the armistice on November 11, prevented the fall of the fortress city to the Americans. In this final battle, the Americans helped bring an end to the war but suffered more than half their casualties during the war, a total of 117,000. The French suffered 70,000, and the Germans 90,000.

OUTCOME: On "the 11th hour of the 11th day of the 11th month," a cease-fire was in place upon the signing of the armistice in a rail car in Compiègne. The war ended with the Allies suffering 22,477,500 casualties and the Central Powers 16,403,000. The world was in disarray and deeply wounded. Shortly afterward, the following occurred: the Spanish Flu spread and killed millions, a generation found itself displaced, food sources were destroyed, communism was on the rise, colonies revolted, the Great Depression was just around the corner, and nationalism was on fire. In hindsight, the Treaty of Versailles treated Germany unfairly, and Adolf Hitler stepped forward to "save" the country and seek revenge. Eastern Europe was fighting the Reds, Japan was invading China, the Ottoman Empire was in steep decline, creating a vacuum in the Middle East, but America was on the rise as a new Super Power.

WHO? August 8-November 11, the Allies suffered the following casualties: France 531,000, Britain 411,636, and America, 127,000. The Germans suffered 785,733 casualties. Of the Germans captured, 188,700 were by the British powers, 139,000 by the French, 44,142 by the Americans, and 14,500 by the Belgians.

WHY AT ALL THOSE LOCATIONS? Commander Foch attacked on a broad front, taking advantage of the fact that Germany had to reorganize itself after its 5 major offenses. The Germans couldn't repel the newly arrived American forces or the reallocated British forces that had come from the Middle East and Italy.

TECHNOLOGY/TACTICS: In tactics, mobile warfare was re-invented through the use of combined arms, including the use of improved tanks to break the stalemate and overrun the strong German defenses. Automatic weapons became more widespread and mobile. The increase in firepower and mobility allowed for violent breakthroughs and "progress" in the war.

PLACES TO SEE:

BATTLE OF SAINT QUENTIN AND SECOND BATTLE OF THE SOMME & AMIENS:

1. **BELLICOURT AMERICAN MONUMENT (ACROSS THE CANAL):** This commemorates the 90,000 American troops who fought with the British and Australians in this sector to capture a section of the Hindenburg Line. The monument is above the canal that was part of the Hindenburg Line, which the Allies broke in September, 1918.
   N 49 58.499 E 3 13.891 (49.974983, 3.231517)

2. **AMERICAN CEMETERY OF THE SOMME (BONY):** In the farmlands of Picardy not far from the Bellicourt Monument is a cemetery that holds 1,844 American graves. It's located on the site of the strongest German position on the Hindenburg Line broken by the Allied forces. The town itself was not cleared until October 1, after several attempts.
   N 49 59.126 E 3 12.958 (49.985433, 3.215967)

3. **RIQUEVAL FARM BRIDGE:** The 6th Battalion North Staffordshire Regiment captured this all-important bridge across the canal. On the east side of the bridge is a partially destroyed pillbox.
   N 49 56.392 E 3 14.523 (49.939867, 3.24205)

4. **BELLICOURT BRITISH CEMETERY:** On the western edge of town is a British cemetery for soldiers who were killed in action in the area (not necessarily during this battle).
N 49 57.678 E 3 13.727 (49.9613, 3.228783)

5. **LA BARAQUE BRITISH CEMETERY AND 46TH ID MEMORIAL:** Just south of this small cemetery is a memorial to the 46th British Division that fought in the battle for St. Quentin.
N 49 55.666 E 3 14.985 (49.927767, 3.24975)

6. **AUSTRALIAN 4TH DIVISION MONUMENT:** This memorial is on a dirt farm road and serves as a remembrance of the Australians who fought in this sector.
N 49 55.837 E 3 13.604 (49.930617, 3.226733)

7. **FRENCH ST. QUENTIN WAR MEMORIALS NEXT TO THE CANAL:** In St. Quentin is a large memorial to the French and the battles they waged to capture this sector from the Germans.
N 49 50.498 E 3 17.882 (49.841633, 3.298033)

7a. **THE KNOLL:** Overlooking Bony from the north is a German strongpoint, "the knoll," that inflicted hundreds of casualties until finally taken by the Allies. In fact it was the main contributor to 995 casualties inflicted on a regiment of the 27th ID in one day. Currently, it is part of a private farm.
N 49 59.411 E 3 12.116 (49.990183, 3.201935)

BATTLE OF DROCOURT QUEANT LINE, AUGUST, 1918:

7b. **CROWS NEST, CANADIAN MEMORIAL, HENDRECOURT:** Located between Arras and Baupme, the Canadian Corps launched an attack on the German Droucourt Queant Line in late August, 1918. After fierce fighting, they captured the dominant position nicknamed the Crow's Nest. Today, a brick monument with a Canadian flag stands alone in a field at the foot of it the Crow's Nest.
N 50 13.150 E 2 57.273 (50.219161, 2.954557)

7c. **UPTON WOOD CEMETERY:** Across the fields in the tree line from which the Canadians attacked is this small cemetery with 226 graves. The area was captured by the 1st Canadian Division on August 30, 1918.
N 50 13.463 E 2 57.183 (50.224386, 2.953051)

SAINT MIHIEL OFFENSIVE:

8. **AMERICAN CEMETERY, SAINT-MIHIEL:** This location holds the graves of more than 4,000 soldiers who died in the Saint-Mihiel offensive.
N 48 57.275 E 5 51.099 (48.954583, 5.85165)

9. **AMERICAN MONTSEC MEMORIAL:** Nineteen km southwest of the Saint-Mihiel Cemetery is an impressive American memorial on top of the dominating Montsec hill. The Americans captured the hill after intense artillery fire and a smokescreen.
N 48 53.403 E 5 42.809 (48.89005, 5.713483)

10. **1ST DIVISION MEMORIAL, VIGNUELLES:** This is a memorial commemorating the US 1st ID; it's in the approximate location where the 26th and 1st Divisions linked up ending the battle of St. Mihiel.
N 48 58.708 E 5 42.909 (48.978467, 5.71515)

11. **LES ESPARGES FRENCH CEMETERY, MEMORIALS, TRENCHES, AND BUNKERS:** At this location is the French cemetery du Trottoir, with 2,960 graves and 85 unknown soldiers. The immediate area contains walking paths to other sites and memorials, including well-preserved German trenches, mine craters, and concrete bunkers. Another memorial to note is the monument to the 106th and 132nd French units. Take the path of the Monument du Coc that runs through the woods and past trenches and bunkers. This coordinate will take you to the middle of the hill, and from this location, you can follow the signs and paths.
N 49 03.843 E 5 36.593 (49.06405, 5.609883)

12. **GERMAN CEMETERY, THIAUMONT:** This cemetery is located near the American Saint-Mihiel Cemetery; more than 8,700 Germans are buried here, plus an additional 2,900 in a mass grave.
N 48 56.754 E 5 52.649 (48.9459, 5.877483)

13. **CROIX DES CARMES, TRENCH LINE/BUNKERS AND FEY:** Located in the woods at the battle's front lines is a cross commemorating the fallen. The site includes information boards, and trench lines and shell craters cover the area. Also nearby are the ruins of Fey, a community that like many towns along the front line, was utterly destroyed and never rebuilt.
N 48 54.981 E 5 59.331 (48.91635, 5.98885)

14. **GERMAN BUNKER:** At this site sits a German bunker that watched over the road (now D3).
N 48 55.753 E 5 54.102 (48.929217, 5.9017)

15. **REMENAUVILLE CHAPEL:** Just off D3 near the bunker is a rebuilt chapel, all that remains of a town of 130 people.
N 48 54.562 E 5 55.768 (48.909367, 5.929467)

THE CHAMPAGNE OFFENSIVE:

16. **AMERICAN SOMMEPY-TAHURE TOWER MEMORIAL, MONT BLANC:** As part of the Champagne counter offensive, the American 2nd ID captured this dominant ridge, called Mont Blanc on October 3-4. Hill 210 is just to the southeast from the memorial, covered by a patch of trees. At this location is an impressive American Battle Monuments Commission tower, surrounded by battle scars and information boards. On reaching the memorial, one can immediately see the geographic importance of the location. The 36th ID replaced the 2nd ID on October 7 and continued the offensive to drive the Germans back north to the Aisne River.
N 49 17.040E 4 32.064 (49.284, 4.5344)

17. **MONUMENT OSSUAIRE DE LA FERME DE NAVARIN (BATTLE OF CHAMPAGNE):** This is the burial site of 10,000 French soldiers who fought in the Champagne campaigns, just minutes from the Sommepy Tower. It was lost in the German July 15, 1918 attack and then recaptured on September 26 by the French 22nd ID during the Champagne Battle.
N 49 13.117 E 4 32.535 (49.218624, 4.542257)

18. **GERMAN BLOCKHOUSE:** One of the remaining German bunkers in the area, just yards from Navarin.
N 49 13.327 E 4 32.653 (49.222117, 4.544217)

19. **FRENCH CEMETERY MONT DE LA GARDE AND GERMAN MARNE CEMETERY:** Located just south of Navarin are a very large French cemetery and a German one that sit side by side.
N 49 11.243 E 4 32.259 (49.187383, 4.53765)

THE MEUSE ARGONNE OFFENSIVE:

20. **LOST BATTALION MONUMENT:** This a marker located on Apremont-Binarville Road. Just down the hill and into the forest are the remains of foxholes and trenches. A German cemetery is also nearby. This is the location where Maj. Charles White Whittlesey held together the remnants of his battalion while surrounded for more than 5 days. Whittlesey was awarded the Medal of Honor for his leadership and heroism. His citation reads: "Although cut off for five days from the remainder of his division, Major Whittlesey maintained his position, which he had reached under orders received for an advance, and held his command, consisting originally of 46 officers and men of the 308th Infantry and of Company K of the 307th Infantry, together in the face of superior numbers of the enemy during the five days. Major Whittlesey and his command were thus cut off, and no rations or other supplies reached him, in spite of determined efforts which were made by his division. On the 4th day Major Whittlesey received from the enemy a written proposition to surrender, which he treated with contempt, although he was at the time out of rations and had suffered a loss of about 50 percent in killed and wounded of his command and was surrounded by the enemy."
N 49 15.072 E 4 54.831 (49.2512, 4.91385)

21. **LOST BATTALION MEMORIAL:** Another memorial to the Lost Battalion a little farther up the road.
N 49 15.054 E 4 54.380 (49.2509, 4.906333)

22. **AMERICAN MEUSE ARGONNE CEMETERY:** This is an extremely impressive memorial located in the hills where the Americans experienced some of the most-intense combat of the war. The strongest section of the Hindenburg Line in this region ran behind the cemetery's chapel. Despite intense German artillery bombardment and machine-gun fire, the US 5th Division captured the ridge on October 14, 1918. It is the largest American cemetery in Europe, and contains 14,246 graves.
N 49 20.046 E 5 05.459 (49.3341, 5.090983)

23. **MONTFAUCON AMERICAN MONUMENT:** At this location the Germans built a strong fortress on the remains of an old abbey. Today, a 200-foot-high American monument sits at the center of where the fighting took place, next to the abbey ruins and bunker remains. You can climb the steps to see the entire battle area. A very impressive memorial. Take note of several German bunkers, including the one built into the abbey remains.
N 49 16.333 E 5 08.500 (49.272217, 5.141667)

24. **ROMAGNE 14-18 MUSEUM:** This is a small, privately owned museum that offers a café. It's located just outside the Meuse Argonne Cemetery and is definitely worth a stop.
N 49 19.930 E 5 05.051 (49.332167, 5.084183)

25. **SERGEANT YORK HISTORIC TRAIL:** The first sign for this trail is at the intersection of Rue des Goulets and D4 in Chatel-Chéhéry. It is a beautiful path built by the French and Americans, and offers explanation boards that trace Sgt. York's feats of heroism. It is not far from the lost battalion location.
N 49 17.008 E 4 57.248 (49.283467, 4.954133)

26. **MISSOURI MONUMENT:** Most of the 35th Division was from Missouri or Kansas. This is a beautiful memorial to their sacrifices in the Argonne Forest.
N 49 13.491 E 5 03.368 (49.22485, 5.056133)

27. **PENNSYLVANIA MONUMENT, VARRENES:** This is a very impressive monument to the battles fought by Pennsylvania divisions in the Argonne.
N 49 13.539 E 5 01.944 (49.22565, 5.0324)

28. **MUSÉE ARGONNE, VARRENES:** This museum is small, but very well done. It contains materiel and displays about both World Wars.
N 49 13.495 E 5 1.893 (49.224917, 5.03155)

29. **CÔTE DE CHÂTILLON:** A ridge and town captured on October 17 by the American 84th Brigade of the 42nd Division commanded by Col. (later Brigadier General) Douglas MacArthur. The regimental commander, Maj. "Wild Bill" Donovan received the Medal of Honor for his bravery during the assault. His citation reads, "Lt. Col. Donovan personally led the assaulting wave in an attack upon a very strongly organized position, and when our troops were suffering heavy casualties he encouraged all near him by his example, moving among his men in exposed positions, reorganizing decimated platoons, and accompanying them forward in attacks. When he was wounded in the leg by machine-gun bullets, he refused to be evacuated and continued with his unit until it withdrew to a less exposed position." The woods contain the remains of trenches and bunkers. This coordinate is for a small parking spot off D123.
N 49 19.768 E 5 02.906 (49.329467, 5.048433)

30. **GERMAN CEMETERY, ROMAGNE:** One of many German cemeteries in the region, situated in a beautiful setting under tall trees and with well-kept memorials.
N 49 20.055 E 5 04.940 (49.33425, 5.082333)

31. **BUTTE VAUQUOIS MONUMENTS AND CRATERS:** A very impressive park where the French fought the Germans in 1915 and the Americans faced the Germans in 1918. It contains various remains of the battles, including bunkers, craters and trenches.
N 49 12.245 E 5 04.081 (49.204083, 5.068017)

32. **HQ CROWN PRINCE:** These woods contain the remains of German bunkers, hardened buildings, and trench lines. They were captured by the US 77th ID on September 28.
N 49 12.846 E 4 59.507 (49.2141, 4.991783)

OTHER LOCATIONS RELATED TO THE END OF WW1:

33. **ARMISTICE LOCATION, RETHONDES CLAIRIÈRE DE L' ARMISTICE, COMPIEGNE:** This is where the WW1 armistice was signed on November 11, 1918. After Germany invaded France in WW2, Hitler had the French sign their surrender in the same rail car. During WW2, the rail car was moved to Germany to protect it, but as the war progressed, it was blown up by the Nazis in 1945. Today, there is a replica rail car and small museum to visit within the city park.
N 49 25.633 E 2 54.383 (49.427217, 2.906383)

34. **PARIS AMERICAN CEMETERY WW1, SURESNES CEMETERY:** Located on the west side of Paris, this cemetery contains the graves of more than 1,500 men who died in WW1 and more than 20 who died in WW2.
N 48 52.315 E 2 13.124 (48.871917, 2.218733)

100 DAYS OFFENSIVE, AUGUST 8 TO NOVEMBER 11, 1918

MAP 25

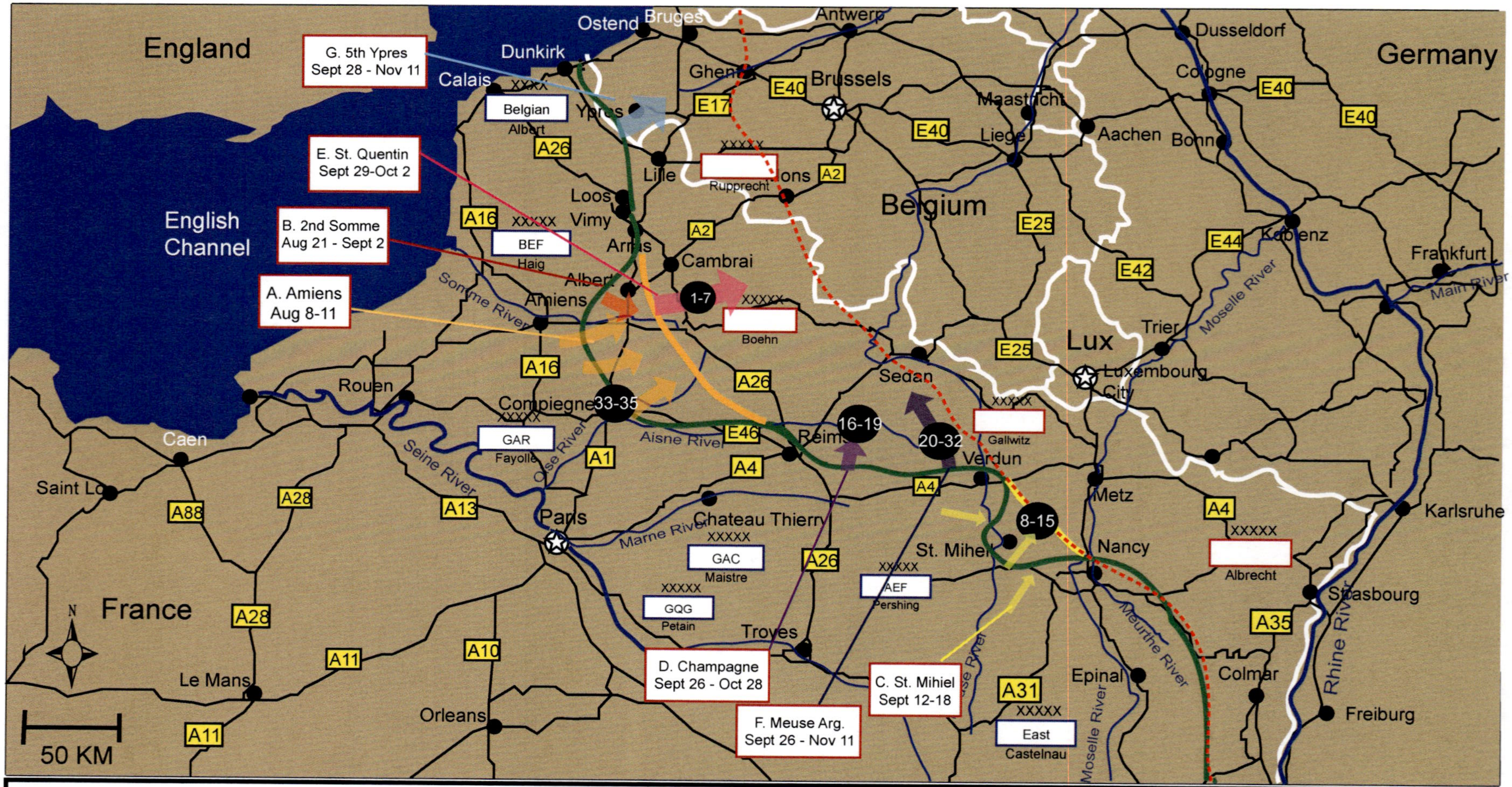

Movements	Results	Places to See
Situation August 8, 1918		
A. Amiens Offensive, Aug 8-11, the British and French launched a surpise attack on the Germans resulting in large gains		
B. Second Battle of the Somme, Aug 21- Sept 2, the British continued the attack to push the Germans back to the Spring lines		1-7
C. St. Mihiel Offensive results, Sept 12-16, the Americans, reduced the St. Mihiel salient to prepare for the big offensive		8-15
D. Champagne Offensive, Sept 26- Oct 28, the Americans struggled to advance to flank the Argonne Forest from the west		16-19
E. Battle of St. Quentin Canal, Sept 29 - Oct 2, an assault on the St. Quentin Tunnel & penetrated the Hindenburg Line		1-7
F. Meuse Argonne Offensive, Sept 26 - Nov 11, part of the "Grand Offensive" the objective was to cut the rail lines up to Sedan		20-32
G. 5th Battle of Ypres, Sept 28 - Nov 11, 1918		See Ypres Chapter
Situation November 11, 1918 (Places to see are Compiegne and unrelated but a great spot, southern front sites)		33-34

ST. MIHIEL OFFENSIVE, SEPTEMBER 12–16, 1918

MAP 26

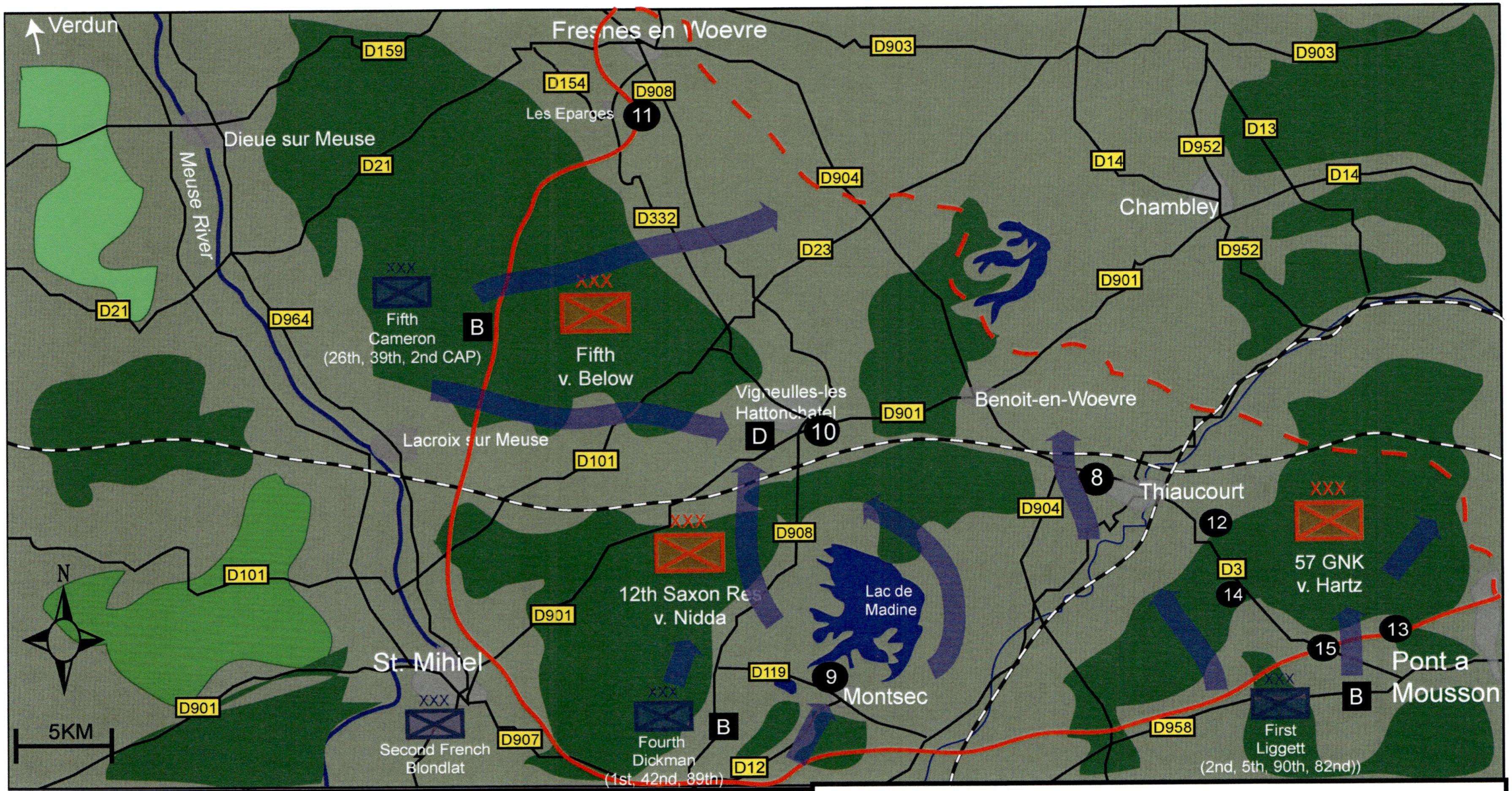

Places to See:

- 8 American Cemetery St. Mihiel
- 9 American Memorial, Montsec
- 10 1st Division Memorial
- 11 Les Eparges Cemetery, Bunker, Memorials
- 12 German Cemetery Thiaucourt
- 13 Croix Carmes, Fey Memorial, trenches, etc.
- 14 German Bunker, side of road.
- 15 Remenauville Chapel/memorial

MOVEMENTS:

A. At 0100 on September 12, 3,000 Allied guns fired rounds of HE and smoke for 3 hours.

B. At 0500 Renault tanks and aircraft began their attack.

C. Right behind the tanks and aircraft, the soldiers advanced into the thick terrain.

D. By dawn on September 13, the US 26th Division and 1st Division met in Vigneulles. The salient had been almost entirely closed in one day, mop up operations continued for another 3 days.

September 12 frontline

September 16 frontline

BATTLE OF ST. QUENTIN CANAL, SEPTEMBER 29 TO OCTOBER 2, 1918

MAP 27

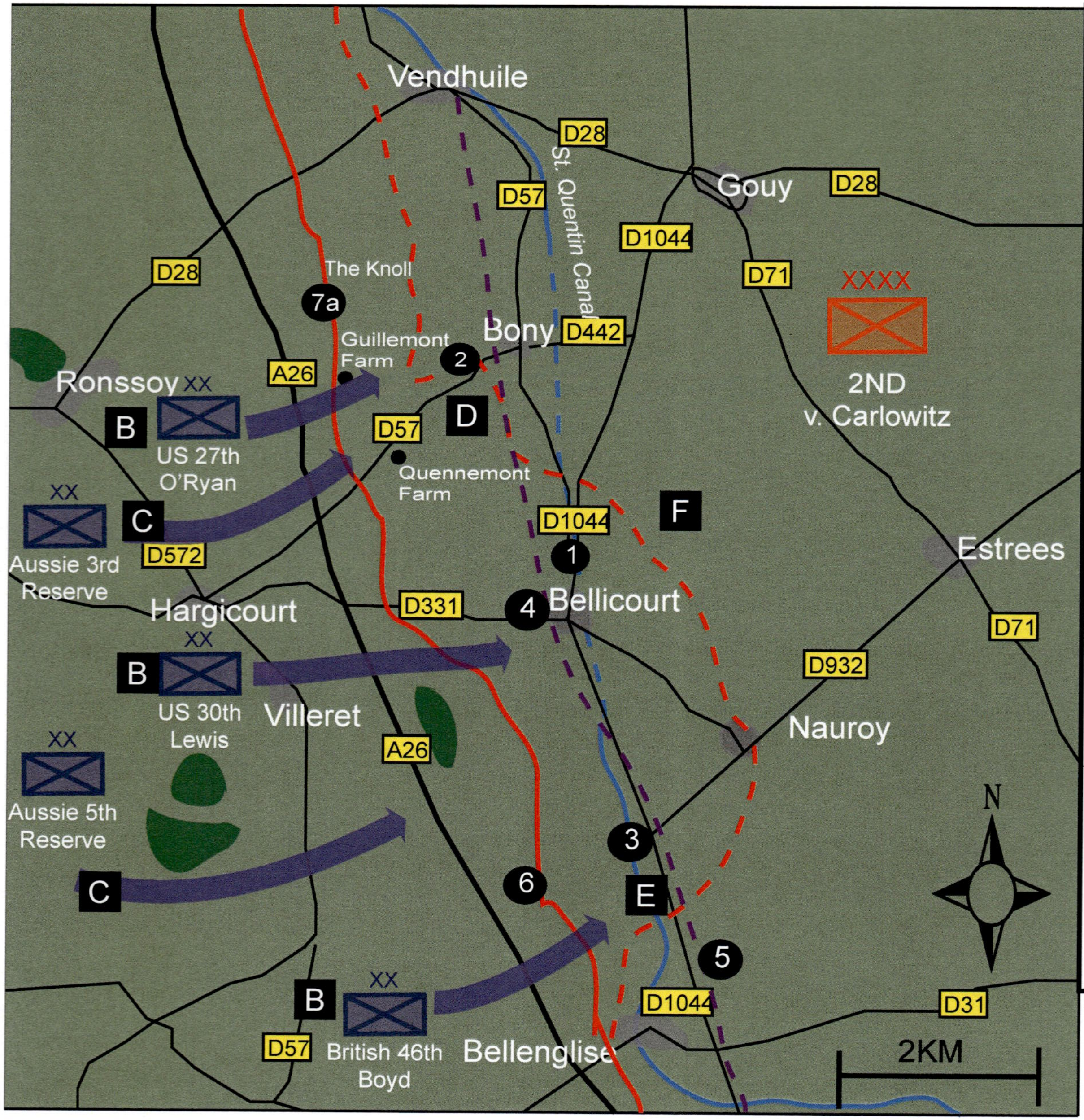

Places to See:

1. American Memorial Bellicourt
2. American Cemetery Somme (Bony)
3. Riqueval Farm Bridge
4. Bellicourt British Cemetery
5. La Baraque British Cemetery
6. Australian 4th Division Monument
7. St. Quentin French Memorial (not on map)

7a. The German Defense, The Knoll

MOVEMENTS:

A. Sept 26-27 Allies fired 1,650 cannons with HE and mustard gas. And units skirmished.
B. Sept 29 at 0555 in fog the 3 divisions attacked.
C. Australian 5th and 3rd Divisions entered behind the American 30th and 27th Divisions because the Germans came out of the tunnels behind the Americans. The US 27th ID was decimated.
D. By noon, the Americans and Aussies captured strong points such as "the Knoll" in the area of the current cemetery in Bony. Bony was cleared by September 30.
E. In the south, the 46th crossed the canal by 0830.
F. By the end of October 2, the 4th Army had penetrated 6,000 yards of the Hindenburg Line at a width of 10,000 yards.

Allied Line September 29
Allied Line September 30
Hindenburg Line Main Defenses

BATTLE OF CHAMPAGNE, 1918

MAP 28

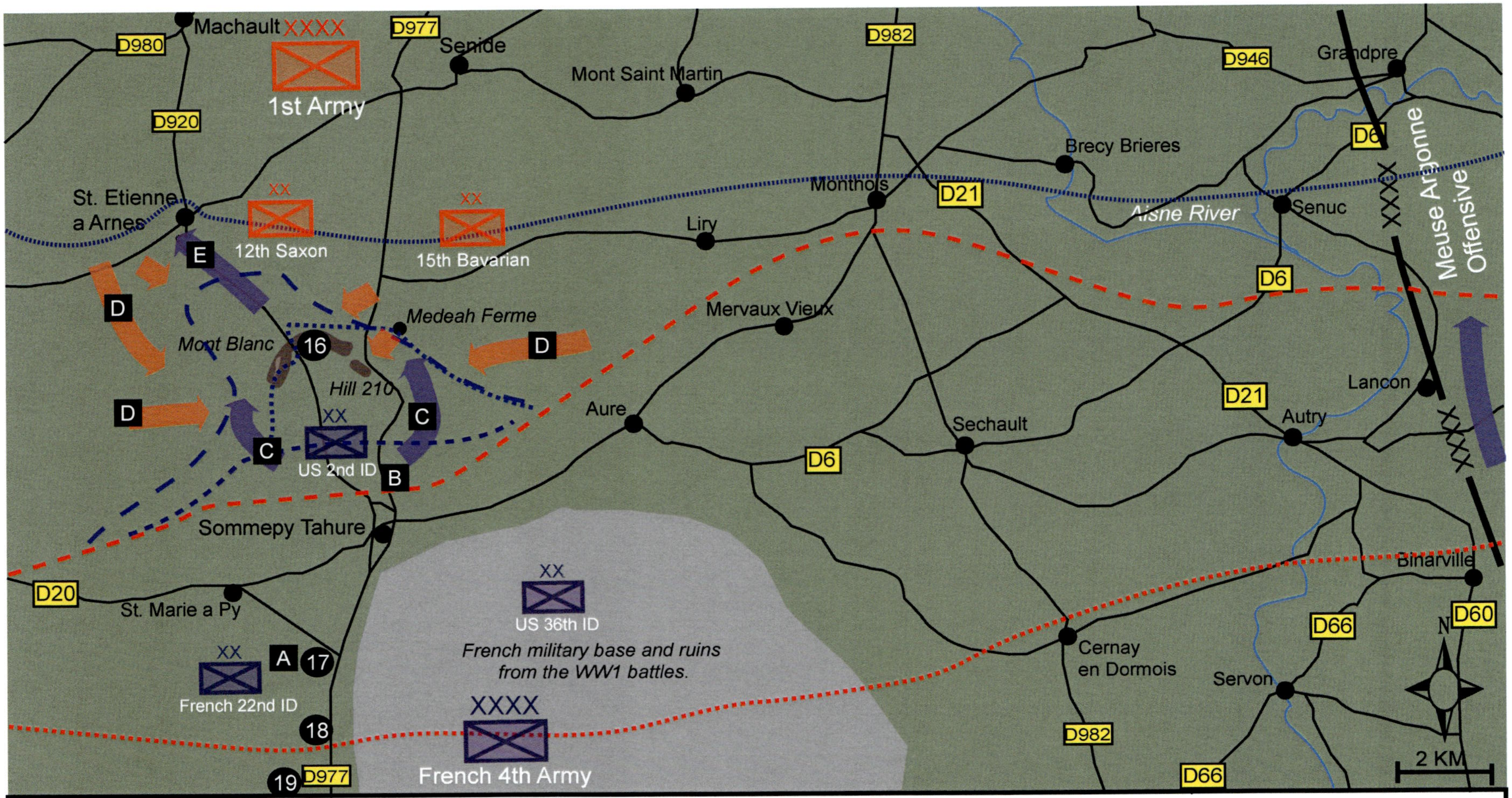

Movements (October 2 - 9, 1918)

A. After the French 4th Army captured Navarin and Sommepy, they could not muster enough strength to overcome Mont Blanc.

B. The American 2nd ID arrived on October 2 and captured additional ground.

C. On October 3, defeating a German spoiling attack, the Americans used a pincer movement to capture parts of Mont Blanc and Medeah Farm.

D. On October 4, the Germans attacked in with reinforcements without success.

E. On October 5, elements of the 2nd ID and French attempted to capture St. Etienne but were repulsed.

F. On October 7, the 2nd ID was replaced by the 36th ID.

G. On October 9, with one French division on either flank, the 36th ID captured St. Etienne. (lines not shown)

Places to See:

16 Mont Blanc American Memorial, Hill 214

17 Navarin Ossuaire and Memorial

18 German pillbox

19 French and German Cemeteries

Sept 26, Front

October 1, Front

October 2, Front

October 3, Front

October 4 Front

October 9 Front

MEUSE – ARGONNE OFFENSIVE, SEPTEMBER 26 TO NOVEMBER 11, 1918

MAP 29

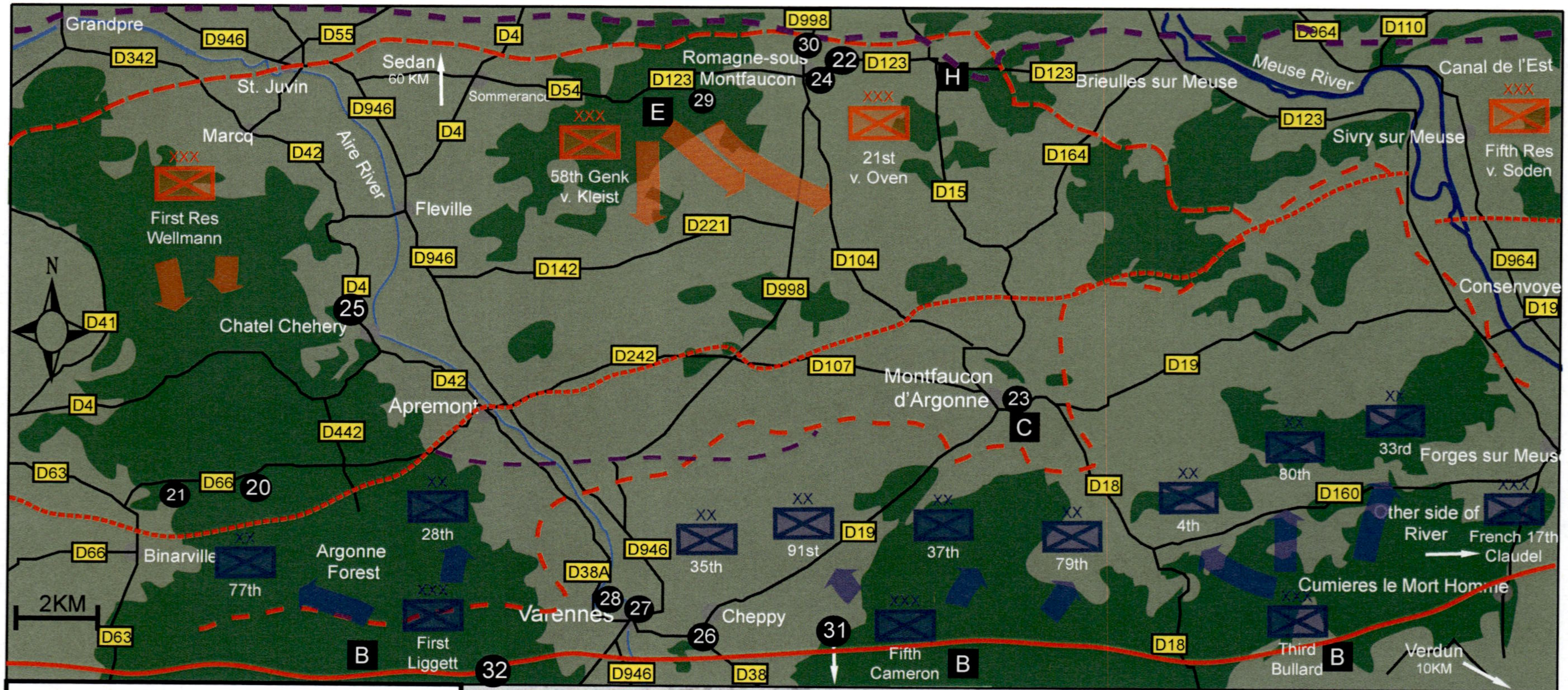

Photo far right is a German dugout in the Argonne and photo center is American artillery on the attack, wearing their gas masks.

MEUSE – ARGONNE OFFENSIVE, SEPTEMBER 26 TO NOVEMBER 11, 1918

MAP 29

MOVEMENTS:

A. 0230 on September 26, the Allies opened up with 2700 guns for a 3 hr artillery barrage.
B. Next the divisions advanced under a rolling barrage.
C. Montfaucon was captured at noon on Sept 27.
D. Sept 29, the US was on the defensive from flanking fire and strong points.
E. September 30, the Germans sent in seven fresh divisions.
F. October 1, the US and French stopped to rest and dig in.
G. October 4, the Americans renewed their attack. In their attack one of the 77th Division's Battalions became surrounded as they out paced their flanks. For 6 days they fought off German attacks.
H. October 14, the main Hindenburg Line was under attack.
I. Nov 1, after a 2 hour barrage, the Germans withdrew over the Meuse.
J. Nov 5, US 5th Division made a bridge head over the Meuse at Dun sur Meuse.
K. Nov 7, US I Corps was looking over Sedan from the south.

Places to See:

20 Lost Battalion Marker
21 Lost Battalion Memorial
22 American Meuse Argonne Cemetery
23 American Monument Montfaucon
24 Romagne 14-18 Museum
25 Sergeant York Historic Trail
26 Missouri Monument
27 Pennsylvania Monument
28 Musee Argonne Varrenes
29 Cote Chatillon
30 German Cemetery Romagne
31 Vauquois Monuments and Craters
32 HQ Crown Prince

Painting top, is John Singer Sargent's "Gassed", completed in March 1919, he was with the British in July 1918 and the American troops near Ypres. This depicts soldiers walking to an aid station after being gassed by Mustard gas The photo above is of American troops, advancing in what appears to be a soldier getting gassed, while the others have their gas masks on. (NARA)

Left, the Somme American Cemetery, part of Saint Quentin Canal Battlefield, the photo right just south of the cemetery is the American Bellicourt Monument over the St. Quentin Canal, where they along with British and Australians broke the German line.

Left: St. Mihiel Cemetery is located south of Metz and marks the sacrifices of the American soldiers who fought during the St. Mihiel Offensive of September 1918.

Above left, is the impressive American Montsec Memorial dominating the area. Photo right is a picture taken from the top looking out northward.

Above left is the memorial in Varennes to the Pennsylvannia soldiers that fought in WW1, it is located on the front line of the Meuse Argonne Offensive. Photo, top right is of the American memorial at Sommepy Tahure.

Photo left is of the Meuse Argonne American Cemetery. It is the largest American Cemetery in Europe. It is located on one of the strongest areas of the Hindenburg Line which was captured in October 1918. In the same town is a museum and cafe and the below left German cemetery.

Left photo, only 6KM south of the Meuse Argonne Cemetery is the Montfaucon American Monument. It was built over the ruins of the battle for a strong German position, located next to an abbey. It is an amazing view to climb to the top.

Left is a marker to the location of the Lost Battalion. The battalion was largely saved and kept intact by the leadership of Major Charles Whittlesey, who was awarded the Medal of Honor for his heroism. (NARA)

In Chatel Chehery, French and American volunteers built a trail to retrace the heroism of Sergeant Alvin York (photo right) (NARA). It is not far from the lost battalion location and well worth the stop for the short trail. We also had the pleasure of meeting the mayor of the town who showed us the start of the trail. (center photo)

FRENCH SOUTHERN FRONT

1914-1918

SUMMARY: South of Metz to the Swiss border along the Vosges Mountains, the French and Germans fought over difficult terrain and made few gains. The land was then German terrain on the western side of the Rhine River, since the Germans had taken it over in 1871 after defeating the French in the brief Franco-Prussian War.

During WW1, when the Germans attacked the northern route to France, the French counterattacked in Alsace through the Vosges. The objective of the attack was to capture a major route into Germany. By August 26, however, reinforcements were needed near Paris, and the Alsace Army was dissolved. The Germans also reduced their forces and moved the 7th Army to Aisne Sector. Small attacks ensued for the following year, such as a French Christmas Day attack to capture Hartmannswillerkopf, which dominates the valley and plains below it. After fruitless attempts by both countries to attack each other, the front largely stabilized and remained stagnant. To note, Frenchman Andre Laffargue who experienced the raids and fighting in the Vosges wrote a document recommending the French Army use heavily armed small groups to breakthrough German lines, bypass strong points, attack the rear areas and let others mop up. Unfortunately at that time, his recommendations were largely overlooked. However, the Germans instead began developing their storm trooper tactics before the French. It is unique to visit because it is so different from the other battlefields. The memorials and museums are worth the trip, while also seeing WW2 memorials and tasting the Alsace wine and food!

PLACES TO SEE:

1. **BATHELEMONT AMERICAN SOLDIER MEMORIAL:** At this location is a memorial to the first shot fired by the Americans on French soil, which was fired by Sergeant Arch Battery C of the 6th Artillery Regiment on October 23, 1917. The memorial is also here to remember the first American soldiers killed on French soil, which occurred on November 3, 1917 when the Germans attacked a location called Le Haut des Ruelles or Gyps Hill. The American soldiers killed were Corporal J.B. Gresham, Private M.D. Hay and Private T.F. Enright. They were part of F Company of the 16th Regiment. The memorial was created by the French in the surrounding area.
   N 48 41.433 E 6 31.170 (48.690548, 6.519495)

2. **MUSÉE MÉMORIAL DU LINGE:** This site includes a very well-done museum and trenches to walk among. In my opinion, the Linge is one of the best museums and sites dedicated to WW1. It also includes German and French cemeteries.
N 48 05.002 E 7 08.398 (48.083367, 7.139967)

3. **GERMAN CEMETERY, DU LINGE HOHROD:** At this location, near the du Linge memorial is a German WW1 cemetery and memorial with the remains of 2,460 German soldiers, of which 1,518 are in graves and the remaining 942 are in a ossuary.
N 48 04.750 E 7 08.629 (48.079175, 7.143822)

4. **VIEIL ARMAND BATTLES OF WORLD WAR I, HARTMANNSWILLERKOPF:** This site includes opportunities for hiking, as well as trenches, a museum, and cemeteries. It offers an incredible view of the Alsace Plain. The Germans and French fought over the mountaintop during WW1. Over 30,000 soldiers perished in the fighting in this area.
N 47 51.524 E 7 08.958 (47.858733, 7.1493)

SOUTHERN FRONT, AUGUST 1914 TO NOVEMBER 1918

MAP 30

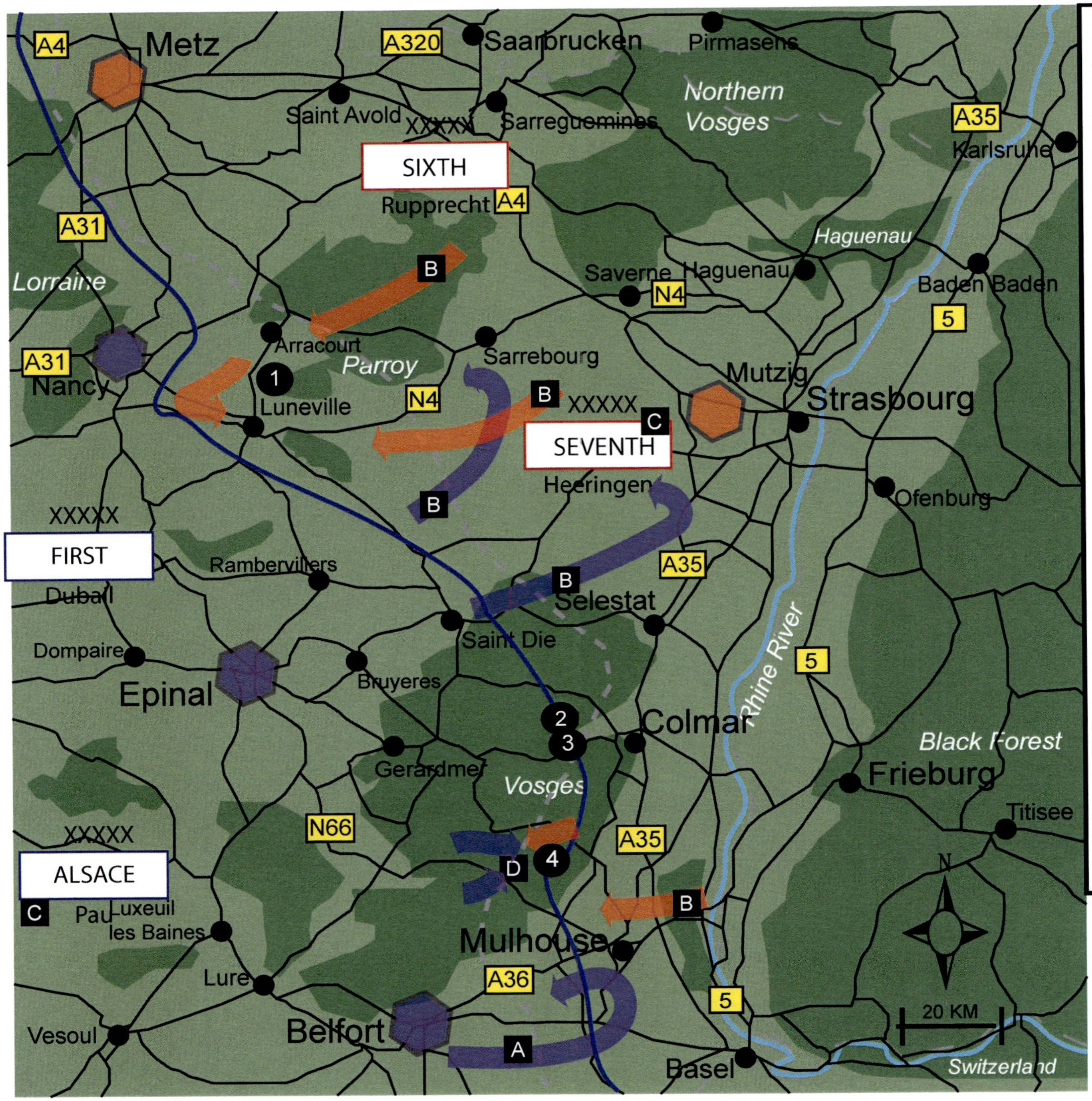

MOVEMENTS:
A. After the Germans attack the French in the norh, the French counterattack in the south near Mulhouse from August 3-10.
B. The attack was unsuccessful, on August 14 to August 26 the French and Germans both attacked and counterattacked eachother with some gains.
C. On August 26-27 the French pulled the Alsace Army back to Paris and the Germans pulled the 7th Army to Aisne.
D. In December 26, 1914 to January 16, 1916 both sides tried to obtain advantage with no success. From then on the line stayed stagnent until 1918.

PLACES TO SEE
1. American memorial, first KIA
2. Memorial and Museum du Linge
3. German Cemetery du Linge Hohrod
4. Vieil Armand Hartmannswillerkopf Mem.

Today's Border
Border in 1871-1918
German Fort
French Fort

Photo left, the American Cemetery Suresnes located just west of Paris. Photo right is in Compiegne, a memorial with a sword through the German Eagel, a mild suggestion on the result of WW1.

Above left, is the museum and memorial park in Compiegne which was the location of the surrender of the Germans in WW1. Photo right is the cemetery at the memorial of Hartmannswillerkopf, a strong hold during WW1, in Alsace, Vosges Mountains.

One of my favorite places to visit was the museum and memorial in the Vosges Mountains, Musee Memorial du Linge. It is really well done and an incredible battlefield that is seldom thought of for WW1. However, it played a very important role stopping any German advances in the south and Alsace sector. The location is excellent and a favorite location to visit for wine, food and hiking.

Above left is the entrance of the very well designed and maintained Flanders Field American Cemetery just outside of Oudenaarde, Belgium. It holds the graves of over 400 American fighters; most who died in the last offensives of the war. Below left is the American memorial in Oudenaarde to the 37th and 91st Infantry Divisions who fought under the King of Belgium. The photo right is of the ornate town hall of Oudenaarde.

Left is the memorial in Brussels, to Edith Cavell a British nurse who assisted Allied soldiers to escape to safe lines and nurse all soldiers, Allied and German. Eventually she was found out by the Germans and sentenced to death, after failed efforts by the Allies to free her, she was executed by the Germans. The memorial is also to the nurse, Marie Depage who was killed when the Lusitania was sunk by a German uboat. She had just left America after raising funds for hospitals.

REFERENCE SECTION

WEAPONS OF WORLD WAR 1 SET I

World War I applied many of the industrial revolution technologies to warfare, creating a leap in new equipment which out paced the tactics and planning of the military leaders. It lead to a massive amount of casualties and impacted the entire world. The following are just some examples of the new types of equipment and technology used in WW1.

Smokeless gunpowder: Prior to WW1, in 1886, the Frenchman Paul Vieille invented smokeless gunpowder called Poudre B. It made small arms much more effective in battle as it removed the artillery and rifle smoke which significantly limited visiibility for the soldiers. In 1887 Alfred Nobel used cordite for smokeless gunpowder which was safer to handle and produce. Due to with less smoke, more accurate and reliable rifles and artillery, the battlefield began to change from linear formations and less colorful uniforms as well. The battles also became more lefthal.

Haber - Bosch Process: In the mid 1800s scientists began to realize that nitrogen was a key compenent for fertilizer and for gunpowder. The world used human waste, animal dung and even grinded bones to fertilize. From the 1820s to 1860s, "guano"(bird and bat excrement) from the Chincha Islands of Peru was shipped to the USA, France and UK until it was all gone. Next saltpeter in the Peru, Atacama Desert was identified to be refined into nitrogen for fertilizer but also gunpowder. Demand skyrocketed, Chile went to war with Peru over the land and eventually won increasing their wealth. However by the late 1800s the demand was exceeding the supply and the west began researching the ability to create nitrogen via the laboratory. In the early 1900s the German scientists, Haber and Bosch had made the most progress. During the begining of WW1 the Allies had cut off the German nitrogen supply from South America, which incentivised the Germans even more. By 1915, the Germans had created the Haber - Bosch Process which created nitrogen from the nitrogen present in the atmosphere. It has been great for fertilizer and supporting a growing population but also terrible that it allowed the Germans to continue fighting in WW1.

Trench warfare was brutal. Pictured above left are examples of clubs used in the trench hand to hand combat. This collection is from the Ingolstadt Military Museum in Germany. Above center, is a 12 gauge "trench gun" or shotgun used by the American forces, yet ironically the Germans declared it barbaric and against the rules of war. (I guess the Germans thought mustard gas was ok though) Also a popular tool in WW1 was the .45 1911 colt automatic pistol. Its design is still cherished today. Below right are some the unique German designed pistols used during WW1. Below left is the Lugar P08, a 9mm pistol, well made, accurate but also complex. To the right is the "broomhandle", 9mm Mauser C96. It was also deployed with a buttstock for more accurate and automatic fire. It was popular for the storm troopers to bring quicker fire power to the fight.

WEAPONS OF WORLD WAR 1 — SET II

Both sides had pistols and bolt action rifles. However, the most destruction as far as fire arms was due to the mass introduction of machine guns. The German Maxim machine gun was used on both sides of the war. And both sides developed similar upgrades. The following are examples of weapons that were made to assist in mobility towards the end of the war and assist the storm troopers on the German side or the troops on the Allied side to break through the lines.

Above left is the Lewis gun, which was actually designed by an American, Colonel Lewis, who was rejected by the US Army Department. He then built a factory in Belgium and moved to England in 1914 where it was massed produced using .303 cal. It has the aluminum air cooling shroud and a 47-97 round "pan magazine". It was used through WW1 and WW2. Above right, the French designed Chauchat, fired a 8mm French round and was the "first" mass produced one operatorlight MG, deployed in WW1.The AEF was issued it and found it to be a terrible MG due to the cheap materiels used and the constant jamming due to the conditions of WW1.

Above left, in September 1918, the Browning Automatic Rifle began to be issued by the AEF & replace the Chauchat. It was a heavy but a sturdy and reliable full auto, single operator light MG that fired a .30-06 round. After the war it became a favorite of the boot leggers and gangsters along with the Thompson SMG. The B.A.R went through some slight changes and was issued as a squad light machine gun through WW2 and even Korea and Vietnam. The draw backs were only the weight and the 20 round box magazine. Above right is the Maxim MG turned into a "portable" storm trooper version. I don't think many would want to run with this one but it provided more mobility and contained a 50-100 round drum magazine fitted on the right side of the MG, firing 7.92x57mm.

Above, is the German made MP18, one of the first real submachine guns designed to assist the Germans in breaking through the trench lines with their stormtrooper tactics. It was only available in limited quantities in early 1918 for the last failed German offenses. Yet the idea was quickly adapted.

Another fearsome weapon that was introduced in WW1 was the flame thrower. The Germans first used the flame thrower at Verdun in 1916 against the unsuspecting French soldiers. It worked as a terror weapon but was very dangerous for the operators due to its range and the soldier became a primary target of the French. Soon the flame thrower was eventually adapted by many of the belligerents and used as standard issue during WW2.

WEAPONS OF WORLD WAR 1 — SET III

The technology continued to be applied against both sides, the following highlights some of the other larger items that began to change warfare for ever.

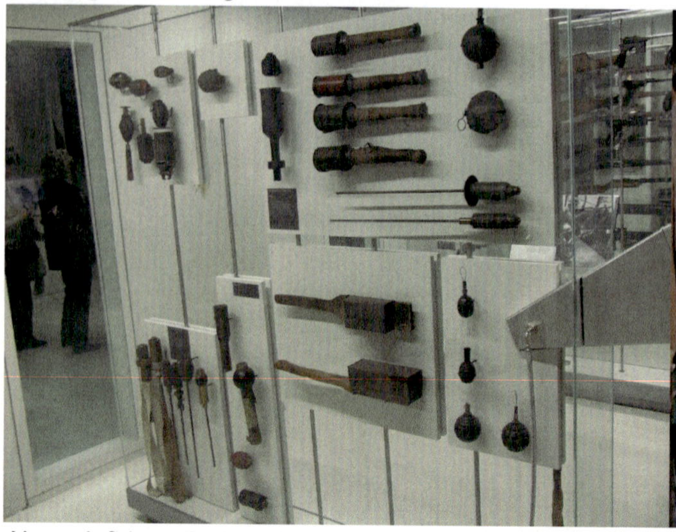

Above left is a display of the various types of grenades used during WW1 on display at the fantasic Meaux Museum. Grenades had not been in much use since prior to the Napoleon Wars due to safety and reliability. Now with new technologies and especially with trench warfare, grenades became essential. Upper right is a display of the various gas masks used during the war to protect against another terrible weapon introduced by mankind during WW1, this display is from the excellent, Museum Passchendaele, Zonnebeke..

Chemical weapons were first used by the Germans in April 1915 with devastating results. Pictured above is a battlefield with the chemical weapon canisters letting the wind carry the gas across to the other lines. Early use relied on the canisters pictured left versus shells and bombs which were still under development. The main issue with the canisters was that you had to wait for the wind and hoped it would not change directions. As the counter measures such as the gas mask were developed, the Germans began using shells that caused vomiting forcing the soldier to take off his mask, then they would shell with mustard or chlorine gas. Photos (BA)

WEAPONS OF WORLD WAR 1 — SET IV

The technology continued to be applied against both sides, the following highlights some of the other larger items that began to change warfare for ever.

Artillery expanded its effectiveness and efficiency during WW1. Pictured above left is the famous and very effective French, 75mm cannon outside the museum in Verdun. It was accurate, light and versatile. The major improvement was that it used hydraulics to keep the gun on target for continuous fire. Photo right is a German trench mortar, a devastating device made in all sizes for use against trench lines, this one is pictured in Ramilles, Belgium.

The German company Krupp made specific cannons to fire against the strong Belgian forts of Liege, Namur and Antwerp. For example, the Germans built twelve 420mm "Big Bertha" siege cannons shown above left. The round weighed 1,800 lbs! It was used with devastating results during the invasion of Belgium in 1914. Not only were the massive howitzers introduced but so were "rail guns" of similar calibur pictured above right with powerful results. (photos from BA left and NARA right photo)

WEAPONS OF WORLD WAR 1

SET V

Far left is the well built French Renault FT-10 tank.It was small, light, reliable and very effective during WW1. Left, is a British tank called the Mark Whippet. It was a light tank used in 1917 - 1930. Both pictures were taken at the Brussels Army Museum.

In the German Munster Panzer Museum they have on display a replica "A7V" tank. It was the only German tank introduced during WW1 and rather unsuccessful due to size, slow speed and only 20 were built. Upper right is probably the most known WW1 tank, the British built MK V "male" tank. These were used at the Somme in 1916 as the first ever tanks in battle. This one is located in the Brussels Army museum. The "male" versions had cannons, the "female" had machine guns.

Not to be forgotten but "barbed wire" was first introduced in WW1. It was invented in the 1880s for ranches in America but soon found the military application during WW1. (NARA)
It caused great problems for mobility and progress. Photo left, from the outstanding French Armor Museum in Samur is the French built Schneider which was somewhat effective but not very reliable due to mechanical breakdowns.

WEAPONS OF WORLD WAR 1 SET VI

Above left is the HMS Dreadnought, it was the first of her type and created the first arms race with all modern countries trying to build the next biggest navy. The Battleship had two major improvements, it had large calibur turrets and it had a steam turbine engine powered by oil or coal giving it great power and speed. Upper right photo, of course was the major development of the German U-Boat, responsible for bringing America into the war. These U-boats are pictured in Kiel, Germany in 1914.(NARA)

As we know, combat aircraft were first used in WW1. Upper left is the famous German "Albatross V" and to the right is the "Camel" aircraft used by the Allies. Both were very effective and reliable. Earlier in the war the Germans introduced the Zepplin aircraft pictured above a German battleship. They were effective initially but soon became easy prey to the ever improving anti-aircraft guns and Allied aircraft. (NARA)

BIBLIOGRAPHY

Anderson, Janice. World War I Witness Accounts, 2010

American Battle Monuments Commission, The 28th Division Summary of Operations in the World War, 1944

Chevallier, Gabriel. Fear. Serpent's Tale, 2011

Griess, Thomas. Atlas for The First World War. Square One Publishers, 2003

Holt, Tonie and Valmai. The Western Front – North. Pen & Sword Military, 2004

Holt, Tonie and Valmai. The Western Front – South. Pen & Sword Military, 2005

Jorgensen, Christer. Great Battles. Parragon, 2007

Junger, Ernst. Storm of Steel. Penguin Books, 2003

Keegan, John. The Face of Battle. Penguin Books, 1976

Keegan, John. The First World War. Vintage Books,1998

Le Maner, Yves. The Two Battles of Bullecourt. Retrieved from www.remembrancetrails-northernfrance.com.

Lewis, Jon. The Mammoth Book of Eyewitness World War I. Carroll and Graf Publishers, 2003

Martin, Edward. The Twenty Eight Infantry Division, Pennsylvania Guard in the World War, 1924

McMeekin, Sean. The Berlin Baghdad Express. Penguin Books, 2010

Mueller, Robert. Fields of War. French Battlefields, 2009

Osborn, Patrick, and Romanych, Marc. The Hindenburg Line. Osprey Books, 2016

Roberts, Andrew. The Storm of War. Penguin Group,2009

Strachan, Hew. The First World War. Penguin Books, 2003

ACKNOWLEDGEMENTS

Similar to the other two books I have written I must thank my family, friends, veterans and fellow history buffs…or war nerds. They all provided in some way energy or inspiration to continue to complete this volume. I am especially thankful for Karina's assistance, guidance and support in continuing to plug away and finish another book. (The little furry ones, Fiona, Sophia and Frederick were also thankful distractions and comic relief for healthy breaks from writing!)

As before also want to thank the many museums, small communities and cemetery caretakers for continuing to remember our fallen heroes and ensure they are memorialized as best possible. A great amount of effort, hours and money has been expended by these small communities and individuals to maintain the memories but especially for the centenary of the Great War. They do an incredible job and my photos and words do not do them justice. I also appreciate having the opportunity to meet and talk with Philippe Gorcynski, the researcher who found the British tank from the battle of Cambrai, in his newly opened museum. Philippe and others like him always provide additional inspiration and set a benchmark for dedication.

I am also very thankful to the North East Public Library and American Legion for their research assistance for the men that fought in World War I from our hometown. Thank you as well to the Army Heritage Education Center (AHEC) of the Army War College for assisting in research of the various units and especially information on the 28th Infantry Division. Similarly greatly appreciate the assistance of the folks at the National Achieves and Records Administration (NARA) who were always very professional and helpful.

In assembling the final product, I would like to thank Anita for her artistic expertise finalizing the cover of the book, Howard for his in depth copy editor talent and Mathew again for his outstanding skills compiling the words, maps and photos into a publishable book.

I look forward to publishing additional books after some more adventures with new and old friends. If you ever have any questions, please contact me on Instagram at @ theroad2war or email me at glxjrd@yahoo.com.

Made in United States
Orlando, FL
04 November 2026

90204700R00131